How to save our world

Paul Arthur van Doorn

How to save our world

The ultimate spiritual society

Uitgeverij
Aspekt 2010

How to save our world
© Paul Arthur van Doorn
© 2010 Uitgeverij ASPEKt
Amersfoortsestraat 27, 3769 AD Soesterberg, The Netherlands
info@uitgeverijaspekt.nl
http://www.uitgeverijaspekt.nl

Typography: Aspekt Graphics
Cover: Aspekt Graphics
Druk: Krips b.v. Meppel

ISBN-13: 978-90-5911-870-6
NUR: 740

For Anne and Maura

(And for everyone who believes in the positive strength of love)

Contents

Preface

Why does someone start writing a book about a subject like this? I have always shown an interest in philosophical topics. Even as a child I felt like the eye in the sky. As if I was not a part of real life, only observing everything going on in the world from above. I tried to understand life from different perspectives. In doing so I found various explanations for these subjects. One of the examples of my inquisitiveness is illustrated by the fact that I wrote an article about "the free will of men" when I was a student in Curriculum Development at the University of Utrecht.

For years I have tried to participate in our modern society always ending up concluding that many things are not righteous or fair; people being treated mean deliberately. This makes me feel uncomfortable about the world I live in[1]. But what can I do? I could not answer the question, which frustrated me. I kept asking myself the question over and over again, why am I here? Is there something I can do to contribute to a better world? After years of searching I got my answer: 'the society on Earth needs to take the next step in order to survive. I will convey this message and how to do so in this book.'

One would like to know where this message came from. From 'God'? From the universe? Did I make it all up? Frankly, I can't tell you where this message came from. I for one think that the answer to this question does not really matter. What I do know is that it makes me feel good and the more I make it work the better I feel.

Meanwhile I came to understand that the universe is much more complex than I would like it to be. Yet, I am still trying to comprehend the greatness of it all.

If and when I am able to understand the universe I might know the reason for writing this book. However, I do not think that a scientific explanation is necessary to understand the reason for writing this publication. I believe that

1 Don't think that I am unhappy, because that's not the case at all. I am living a fantastic life! There are a lot of nice people in my life, I do not lack anything, I am in good health and I am not confronted with suffering or pain. I am very grateful for being so fortunate.

more people will convey a similar message. It is a logical consequence of what is happening here on Earth.

It is good to see that the number of people who achieve a common state of mind increases every day. It signifies that things will really change. I am looking forward to the day this will happen. Then I will understand where this message comes from.

A message may hold various meanings. A message without meaning will reveal its meaning just as well. Therefore, I would ask anyone who reads this book to reflect fairly on the meaning of this message. Does modern society contribute to the common benefit and welfare of the entire population on Earth in the near and distant future? Does society as described in this book offer guarantees of higher quality to the welfare of all and the survival of human life on Earth?

A few years ago I found myself looking into a mental mirror. My mind and my emotions were not balanced. Because of this discrepancy I wanted to understand more of the emotional aspects of my life.

I wondered who I really was and what my main goal in life was about. Many questions came from these contemplations: 'Does God really exist?', 'Is life just about having fun or is there a higher purpose?' and so on. In those days I read many books about Christianity, Islam, Buddhism, near-death-experiences and suchlike.

Processing all this information triggers many more questions. I wanted to answer these questions and sought for the solution in my inner self.

First, it was all dark and vague. As time passed it became clear to me what it was all about.

I went down a narrow path which became a brightly lit highway, illuminated from either side of the road. It made me feel good. For at last I knew what I had to achieve and in which direction I was supposed to go. I did not need a navigation system of any kind, intuition would lead the way.

In early 2003 I realized I had to write a book telling people that generally speaking all religions[2] are similar in meaning and in essence. This also means

2 The term religion is used here in the broad sense. It concerns not just faith from mainstream religions, but also the more general form of spirituality, feelings, thoughts regarding the meaning of life in relation to a particular principle (e.g. a way of life), substance (e.g., reincarnation, life after death) or entity.

that it does not make any sense to argue about detailed information. Organized religions are often being abused intentionally by few to exert power over many.

Many questions about the existence of a 'God' and a life after death are answered by means of near-death-experiences and paranormal phenomena, although not yet scientifically substantiated. In 2007 I found out that a book[3] had been published in which many of these similar ideas had been worked out. The author of that book also draws some conclusions[4], on which I fully agree because I had already come to the same conclusions. This book confirmed everything I had already figured out myself. I did not want to write a similar book, because it would probably add very little.

The foundation for my book about the spiritual society had already been established! It was time to start writing my own story; starting as soon as possible because I do not know how much time I have in this life to finish this 'mission'.

For a thorough understanding of the assumptions about giving meaning to life I recommend the book by Bob Coppes. It will clarify the subject in this book.

Now I understand why I am in this world.

As a child I wanted to be a millionaire. As I grew older I did not understand what went wrong in obtaining a lot of money. However, if I had become a millionaire chances would have been small for me to think about the world the way I do now. I would not have realized that I had to convey the message I am now writing about.

Now I understand why some people obtain material wealth while others live in poverty.

Still, I am free to tell my message. That is how I became aware that I had to tell my tale and make sure that my message would be heard.

My respect for the greatness of the universe has grown. I wish all people the same insights and even more, so that in the future there will be a heaven on Earth, a place where all people live in harmony appreciating and respecting each other.

---

3 Near-death-experiences and world religions, Bob Coppes, ISBN 90-5911-680-1 or look at www.bobcoppes.com.

4 See the Annex.

What applies for me may apply for other people: You have to do what you think is right. Everyone knows right from wrong in their hearts.

People will also be aware of the mind games they play in order to take advantage of other people either for the sake of financial profit or to gain power over others. These and other matters might restrict people to thrive and develop.

The more people are true to themselves and treat others fairly the better the world will become as a place to live in.

I would like to thank everyone who encouraged me to write this book.

I would like to thank my family, friends and colleagues who encouraged me at moments when I could not find the energy or inspiration. They gave me strength by telling me that I was doing the right thing. They assured me that writing this book was for a useful purpose.

My special thanks go to the reviewers Helena de Wolf, Bob van den Heuvel, Vera Breg, my brother Hans and both my parents. They supported me in writing a book accessible to a large audience. And of course to Lucie Evans and Ritchie de Quay without whom this translation would not have been possible.

The fact that more and more people reflect on today's society and try to come up with better solutions for the problems we come across encouraged me to tell this story which you are about to read. We are looking for solutions but not for the cause of our problems. As long as we keep acting like this we are only concerned about symptom control i.e. keeping the system going.

This is not an exciting boy's book or interesting novel. It is a book conveying the message about our spiritual society. Nevertheless, I hope that everyone will enjoy reading this book and that people will be incited into new and positive matters. Pass on the message and it might lead to actual changes.

Introduction

This book is a message to everyone on Earth regardless of race, religion, gender, sexual inclination, political preference or intellectual ability.

Intellectuals already know that this world is not a fair place and that it should be changed into a healthy, fit and improving environment. However, the wealthy position intelligentsia is holding causes a slow down and full stop in solving the problems.

The vast majority which goes without holding such a comfortable position might be able to make a difference if only people were united in demanding for a different world to live in!

It should be a world in which everyone gets a fair chance of happiness. This book outlines what this world may look like and how to achieve that goal. There is no excuse to go on like this. Indeed, the current situation requires changing the world while we still can!

This book is written in fairly simple terms, understandable for everyone.

As a student I was the youngest graduate ever in the Netherlands. Although I was trained to solve problems in an academic way my motto has always been: 'complex problems can best be solved by means of simple, logical answers revealing the core of the question'.

I for one believe that people can only find happiness, if they understand the world they are living in.

The first thing I would like to explain is the notion that we must learn to live in harmony with ourselves, other people, the Earth and the universe.
The size and quantity of raw materials in this world are limited. That implies that we should use a smart system for dealing with that world. That mechanism should be based upon balance not on growth. It should be a system in which everyone gets equal chances and similar opportunities to obtain welfare. It should be a system allowing us to work together in order to accomplish welfare for everyone.

Honesty, respect, openness, happiness, freedom, love and welfare are within everyone's reach and are not just empty promises.

If the world population becomes too large people will be trampled by one another and more wars for natural resources will likely emerge.

In the long term the struggle for natural resources will lead to a situation in which raw materials will only be accessible to a minority of people. These happy few will be the ones possessing all the money and power.

In the spiritual society money and power will become less important. The only power is the universal entity[1] and love is the greatest good we seek.

Most people in our society claim to be happy. People in poor communities claim the same thing. Apparently, prosperity is not the right and only means of measuring happiness and welfare. What might the right measurement for happiness be? Are people really happy or do they deny the fact that there are many reasons for their unhappiness?

Approximately 10% of people in rich Western countries are living in poverty. Many people in the other parts of the world do not have roofs over their heads. They do not have a healthy diet and clean drinking water. Neither do they have access to health care.

Many people are being oppressed, treated as slaves or happen to be victims of violence or war. Others have to deal with illness suffering from misery and death around them.

Logically speaking, there are a great number of people who are not happy at all. But it is nice to know that most people think in a positive way and strive for personal happiness whatever situation they happen to be in. They definitely deserve better than that. They are like you and me. As children they had the same dreams as you and I, hoping for a beautiful and fine future. How different reality is!

At the end of 2005 the entire world still believed in the positive aspects of the capitalist system. Communism experienced a setback because of the disintegration of the Soviet Union. The People's Republic of China remained a communist society until today but capitalist initiatives have been admitted increasingly. Cuba is still a communist country as well but the Castro regime is near to its end. Years of isolation (as a result of the

1 An entity is an existence whereby the status of being is emphasized. Many people consider this greatness to be a divine concept. The essential point is the notion that the universe contains a power connecting each and every one.

U.S. trade blockade) resulted in a low average standard of living. This is certainly not an example to the rest of the world.

It only raises the question 'What did capitalism bring us?' One day I was watching a television documentary on the regime of former President Suharto of Indonesia. The program makers stated that Suharto brought prosperity to his country as well as a substantially higher standard of living. They also claimed that only 15% of the people really benefited from this fortune. This discrepancy prompted me to contemplate on capitalism. It struck me that only a relatively small proportion of the population benefits from the capitalistic system, whereas the majority of mankind is living to a moderate or even poor living standard.

Is capitalism bad altogether? I guess not. Capitalism took humanity a step further. However, all episodes come to an end. Now it is time for a better world. We have learned that current systems have many problems. If we want to continue life on Earth, and really want a better life for everyone, then we need to change, reorganize, and renew.

One of the people who read the concept of this book, asked me: 'Although every life comes to an end eventually – even the world itself will come to an end - your wish to save the world is admirable. But what do you want the world to be saved from?

- Total destruction of the world;
- The destructive influences that nature has upon the world;
- The destructive influences that humanity has upon the world;
- Destruction of mankind itself.'

I needed to think about the answer to this question (for a moment). First, I thought about our ecosystem with regard to its climate and natural environment. These are the things on which mankind depends. Serious disturbances in this system and these elements will have enormous consequences for the world's population. Nevertheless, all the above mentioned answers are true, because it is difficult to predict our future. I for one think that a spiritual society will save many of the good things on our planet. That's one of the strengths of the spiritual society!

There are more and more people who believe in a spiritual world as a solution though it is still a future image. What will this world look like exactly? By the time I started writing my book I saw a TV-commercial showing a district with only detached houses and large, fast cars as if this place was Paradise. Many of you might also have seen the movie 'Pay it forward' in which everyone who receives a good deed is obliged to pass three good deeds forward. Such a concept is mere idealization many people will say.

Is there any reason why such a scenario should not be real? What should a spiritual society look like? That is precisely what I am trying to describe in this book. One may also consider the notion of a spiritual world to be a new way of thinking in order to solve the problems the world is facing today.

In Chapter 1 I describe my vision/view on the history of the world. I will expose many well-known issues in an unorthodox way. The classical views on these widely spread issues are as true as my way of thinking. This chapter will cover a number of key problems in the world today[2].

In Chapter 2 the problem areas of modern society[3] are described. There are many examples of problem areas in Dutch society. Many of these problems are typical for other Western countries as well. In addition, major problems from other countries will also be appointed.

Chapter 3 describes a number of scenarios showing what might happen if we do not accomplish the necessary changes.

In Chapter 4 developments which are already taking place are outlined. I believe they will contribute to other societies and other insights into human life.

A number of critical questions and comments, which will hopefully encourage reflection, have been added to this chapter.

Chapter 5 illustrates the main principles of the spiritual society. In this detailed representation I talk about the way how the spiritual society may function.

---

2 Till mid 2008.

3 This includes all societies on earth. They consist of many sub societies like the Western society or Dutch society.

The chapter also indicates the problems humanity is struggling with might be solved by implementing this new culture.

Chapter 6, represents a possible transformation process from the traditional way of life into a new way. It elaborates on how things should change in order to achieve the spiritual society.

In Chapter 7, the book concludes with a final word advising the reader to disperse for a while, and think about the new insights you have read.

Before the spiritual world functions properly we face the challenge of getting there in the first place. People need to change the way they think to a large extent. They should be willing to give up power and positions. Besides, many institutions and business companies will be superfluous. They will change or disappear entirely.

A new social system offers new possibilities and opportunities but the ones I am referring to are of a different nature than the ones we know in the current social systems.

It is necessary to focus on the upbringing and education of children because that is where accomplishing the spiritual society starts. Different standards and values should be taught. Great political and cultural leaders must take up their responsibilities as role models satisfying the needs for the entire population to fulfill the concept of the spiritual society.

All these changes will take time so we ought to realize continuing the way we do will increase pain and suffering. Epidemic diseases, natural disasters and wars will require us to change our lifestyle. It will eventually lead to the 'survival of the fittest (or richest)'. This will take us back to the level of the uncivilized, prehistoric Neanderthal man.

We should not be proud of such a decline and I do not really think that people are looking for retrogression.

In this book I elaborate on various subjects demonstrating which changes must be made for a spiritual society to work properly. In doing so I will show examples of some of the problems our society is facing today.

I would like to invite readers to respond to the message I am trying to convey. Together we would be improving my ideas which results in co-operation anyone might contribute to. This increases the probability of acceptance and use of the proposed solutions. In doing so we all contribute to a better world making our children feel proud of us.

Moreover, this might evolve into perspectives of a higher quality after this life.

Chapter 1

History of the World in a nutshell

1.1 Introduction

In this world we truly believe in the present power. We are not interested in the past, and think of the future as far away. But the future starts now and is tomorrow. It is to a great extent the result of what we did yesterday.

Scientists like to make graphs based upon accumulated facts extrapolating these facts into the future. Why wouldn't I do so? I graduated from university, so I have got the expertise and capability of following standard scientific methodology.

The main reason for looking into the past is to learn from it for the future. How should we prevent the extrapolation into the future from becoming reality? The world is always changing and in the next chapters it will become clear that change is necessary and inevitable.

It is not my purpose to summarize the entire history of the world. I would like to expand on a number of things and outline a particular development. I will also demonstrate a number of recurring elements in human history.

I will try to represent my personal view on the history of the world. Many events will not be addressed such as scientific discoveries because most of them are not essential for the development of the present situation in which our society has arrived.

1.2 Prehistory

The genesis of the Earth

The world has gone through an enormous evolution. According to the Bible, 'God' created the Earth in seven days. By now we may assume that this is a simplification of what really happened. Furthermore, it will be difficult if not impossible to prove what the development of the Earth was exactly like.

In the earliest times there were no people and if they had existed at all in those former days they would not have mastered the skill of reading and writing. That does not imply for many people that the Bible explains the whole story of the rise of the Universe and our Earth.

Therefore the biblical explanation for the creation of the world is no more than a religious assumption. But maybe there is no need to explain the creation of the Earth in more detail?

Differences between man and woman

The above mentioned story on the origin of the Earth seems to descend from a far distant past which we left behind a long time ago. Still, many characteristics of present day humans are derived from those early days.

Little girls love to play with dolls and simulate caring activities while boys like running, fighting and acting tough. Apart from the psychological factors we also see physical contrasts between men and woman which are due to the evolution of mankind.

Thus, men and women differ in many physical ways for instance eyesight. This variation affects spatial insights. A woman quickly finds what she needs in the fridge while a man rapidly learns how to discover a deer on the horizon. From the beginning of the history of man, women have always taken care of children. Therefore, they were focused on objects in the vicinity, while men had to recognize a prey at great distances while hunting for food.

The brains of men and women differ in composition and development from an evolutionary point of view. Much of this theory is to be found in recent research[1]. However, it does not mean that men are superior over women or vice versa; men and women are different genders of the same species.

Being superior is a relative matter. An elephant might be taught how to ride a bike but it is more logical to teach an elephant to drag trunks. The same applies to men and women. It would be of greater significance to use the strengths of men and women in a righteous way instead of trying to prove at any cost that men and women are able to perform the same tricks. I am good at cooking dinner but I dare to admit that my wife can cook a better meal.

1.3 The first writings

Gilgamesh-epic

Approximately 3500 years ago writing was established for the first time. Because of the skill of putting facts and fiction into words a great deal of life in those days has been revealed to us.

---

1 From the book 'Why men don't listen & women can't read maps' by Allan & Barbara Pease 1998.

One of the oldest and most famous documents from that period is the Gilgamesh epic. One of the themes in this book is a deluge in the area of present Iraq. We should realize that many of those stories are not first-hand because not many people knew the principles of reading and writing. Therefore, many of ancient tales were passed on from generation to generation in an oral tradition. Only the rich and powerful people who had mastered the skill of reading and writing could afford to archive their ideas. That is why most texts were written under censorship. Democracy and freedom were not part of life during those days. Nevertheless, it is remarkable that the Gilgamesh epic matches the description of the Biblical flood.

Bible, Koran and Torah

Another old book that I would like to mention here is the Tanakh, the Hebrew Bible. The first five parts contain the Torah which in turn forms the basis for Judaism. But these five parts form the basis for the old testament of the Christians as well. In addition the Torah is acknowledged in the Koran, the holy book of the Islam, as a document of God. These scriptures depict a period beginning at 1000 B.C. up to the year in which Jesus Christ was born which was the starting point for the rise of the western era. From there on Judaism and early Christianity started to find their own way along the path of religious history.

Thus, it is important to note that these Holy Books tell an important part of history of the Middle East. It is the area from which a number of religions sharing a common background emerged. The followers of these religious movements constitute the largest parts of the world population to the present day.

It is noteworthy that both the Muslims and the Christians believe that written messages have been received straight from the hands of 'God'. In both cases it concerns the Ten Commandments, which Moses received on mount Horeb in the form of stone tablets.

1.4 The first great civilizations

All over the world mankind developed from prehistoric cultures into many new civilizations. This period lasted until 1000 A.D. The best known examples of this dispersion are the Indians from Northern America, the Mayas from South America, the Egyptians from Africa, the Chinese from Asia, the Greeks and the Romans from Europe.

Most of these societies eventually became powerful empires only to come to an end after having reached their apex. Generally speaking these societies were distinguished by small aristocratic groups holding all the power, knowledge and material wealth of the entire community. In addition, these civilizations were characterized by issues such as slavery, human sacrifices, gladiator battles and wars for the rich to increase their own wealth. The Mayas are still admired for their knowledge of the universe, mathematical calculations and precise calendar. Yet, less noble activities (to our eyes) such as human sacrifices were part of their daily life.

The Chinese

The Chinese dynasties were founded around 1000 B.C. The last Chinese emperor was deposed in early twentieth century and Communism was introduced.

A few hundred years before Christ the Chinese wall was built. It is the most famous building in China. It was built to protect the Chinese empire from foreign enemies. This edifice, thousands of miles in length, is called the longest tomb in the world, because constructing such a large structure caused many casualties.

The Romans

At its highlight the city of Rome had turned into a glorious architectural work of art containing luxurious thermal baths, beautiful theatres, markets resembling shopping malls, the celebrated Pantheon, the breathtaking Coliseum and of course the renowned aqueducts. All the glamour and glory were represented as well in Roman business which was from a governmental point of view well organized and contributed largely to the emergence of such a large empire.

The Romans also kept slaves. Wealthy Romans usually owned quite a number of house slaves who executed household chores. Trading slaves was common business in Rome.

After the rise of Christianity, early Christians were persecuted. They were sacrificed for sport in the great amphitheatres where they were devoured by wild animals such as lions, tigers and wild dogs. Men, women and children were killed because of believing in Christ, a non Roman God. The public loved it. The Romans considered the Germans to be barbaric!

Incidentally, at the beginning of the fourth century A.D. the Roman emperor

Constantine converted to the Christian faith after a vision he had in a dream. From that moment on Christians were no longer persecuted by the Romans.

The Egyptians

The Egyptians also had slaves. The pyramids could never have been built without the muscular power of the slaves. There has been a lot of speculation about how the huge stone blocks were made and transported.

The Egyptians applied all sorts of inventive and useful techniques but ultimately it came down to a lot of slaves who had to work very hard. Many lost their lives. One has got to bear that in mind when admiring these buildings.

Technical development gradually took place although these societies were very primitive in nature. The Egyptian culture was primitive in nature particularly when it came down to respecting the lives of other human beings. It's about the struggle for life; survival of the fittest. The Egyptians lived from war to war. Fighting was common business in Egyptian life.

The first payment arrangements

In the early days the concept of money was not part of man's life. People lived off the land and traded by barter. They traded shells, livestock, flour, salt and gold.

Around 700 B.C. the first coins were used as currency. Paper money became a means of the economy much later in history i.e. between 1400 and 1600 A.D.

An important aspect of trading and paying systems is the means of determining the value of goods. Today, this is a matter of supply and demand while in the past different factors played a significant role. Something might have a certain value today whereas that same object has a different price tomorrow. Good examples of deviation in value are new items compared to antique objects, computer games that devaluate within a year, the price for a holiday outside the season or during a school vacation.

1.5 The Middle Ages

The Middle Ages started off in the sixth century when the Roman Empire collapsed. They lasted until the end of the fifteenth century.

This era is characterized by the emergence of agriculture and stock-breeding on a larger scale due to increasing populations. That's when small 'kingdoms' of large land owners originated. The kings of these realms made agreements with small farmers to work on their land. The land owners often lived in castles and had armies of soldiers and knights. They frequently sent tax collectors to the farmers to collect money. Paying taxes resulted in land owners receiving a lot of money. It was easy money earned without having to work hard for it while the farmers often lived in poverty and had to do back breaking labor. The land owners could afford to maintain their armies and go to war with the taxes they collected from farmers.

In medieval times vast territories were conquered because of the invention of gunpowder, imperialism of the clergy took on larger forms. Again, violence, power and wealth were important motives to make war and capture land. Today, these matters are still ongoing while the only thing being different is the era we live in.

The most famous story of knights and kings is undoubtedly the legend of King Arthur and his magical sword Excalibur. He is said to have built a huge Anglo-Saxon kingdom with the aid of Merlin, the renowned wizard. Until this day there is no evidence for the existence of King Arthur and the Knights of the Round Table.

Historical facts show that kings such as William the Great and William the Conqueror (also known as the Duke of Normandy later on to be crowned King of England) were real persons not just legends and myths from the past. This all happened between 1050 and 1100 A.D.

Another interesting historical figure is Joan of Arc who lived between 1400 and 1450 A.D. during the hundred-year war between England and France. At the age of 17 she commanded the French army and led them to the liberation of Orleans. She was taken prisoner by the English and condemned to the stake to be burned alive. She was believed to be a witch.

In the same period many more barbaric rituals were widely spread, such as the drowning of women who were thought to be witches. Other methods of execution were hanging from the gallows, or beheading by means of the guillotine.

Although trade was common business in the Roman Empire, worldwide commerce expanded to a large extend during the Middle Ages. An eminent example is the VOC (United East-Indies Company) which set sail to Asia with regard to obtaining raw materials such as cotton, silk, spices, gold and ivory. Many of these products were a bargain to the VOC. The need to exploit more and more caused further exploration of unknown land. Thus, Australia and America were discovered and set off a new era.

1.6 Africa, America and Australia

Gold and diamonds in South Africa

Between 1500 and 1800 A.D. America and Australia were discovered. South Africa had already been discovered but heavy immigration came later and did not start until diamonds and gold were unearthed. First, a number of wars against the indigenous tribes had to take place. After the Zulus had been conquered they were put to work as slaves in the mining industry.

South Africa is best known for its 'apartheid', legislation based on race which was abolished in the second half of the twentieth century. Nelson Mandela put an end to apartheid but the economic power is still in the hands of the white upper-class.

North America

Perhaps the most famous discovery of all continents is the finding of America, land of opportunity. European people settled in America at the expense of the Native American people which have almost been eradicated.

Some tribes of the first inhabitants dwelled on the Great Plains living off the great herds of buffalos which were decimated by white newcomers just for the sport of hunting.

The Indian nation disagreed on such waste of their natural food supply and went to war against the white hunters. Many Native Americans were slaughtered by the white intruders. The common armory for Indian people was bow and arrow, spears, sticks and stones while the immigrants had guns and cannons. This was not a match for the native people

The Indians were also misled by the white government authorities. Written documents promised eternal rights for Native Americans. The white government often broke these promises soon after having agreed upon them.

The European newcomers depicted Indians as cruel and primitive people.

Actually the indigenous tribes were taught cruel warfare by the white people who were in general much more brutal. 'Destructive white warfare' was not carried out by the Indians. In fact, Indians lived in balance with nature and only killed animals in order to survive. They were always aware of the fragile, natural balance. They had great respect for nature.

The older tribe members were highly respected for their wisdom. It is noteworthy that a number of tribes such as the Sioux, Cheyenne and Apache had a set of rules similar to the ten commandments of Moses and other universal beliefs. Typical of such a regulation is the respect for all life on Earth including the Earth itself. These Native American tribes also recognized a universal essence or 'God'. They called it the Great Spirit. Many people consider the religious beliefs of the Indian people to be paganism. I believe that we can learn a lot from the balanced way of life of these people.

The famous rock band the Eagles wrote a song about the great migration to the western frontier of North America by the European newcomers. The song is called 'The Last Resort' and part of the lyrics state that 'We satisfy our endless needs and justify our bloody deeds in the name of destiny and in the name of God'. This phrase is a critical remark regarding the relation of the Americans with God.

Things became even worse as the new Americans started to transport people from Africa to North America. These African slaves were put to work on the white men's plantations, mainly in the south of the United States. In the course of a relatively short time many millions of slaves were transported from Africa to America. Again power and money were the motives for people to deceive and oppress. Africa was finally colonized entirely and all valuable resources were confiscated completely. The same happened to India.

The de-colonization of colonized countries took place after the Second World War.

Only by then did reconstruction of colonies become a possible way of regaining their national identity.

The American War of Independence is one of the most significant wars ever to have taken place. The independence of the United States was at stake. The Americans got rid of the dominion by the English. This was approximately in the year 1783. As a matter of fact this war was not really about freedom but about disagreements over tax-regulations.

A century later The American Civil War dividing the Northern and Southern states took place. It was North against South for freedom of all people.

The Southern states lost the war and slavery was abolished. That's when true freedom became a nationwide concept in the United States. This happened at the end of the 19-century.

Australia

In Australia the same happened to the Aboriginals who were driven away from their fertile land. They were deceived at negotiations over their land. Only few Aboriginals are still living in the deserts which are places barely endurable for human beings.

1.7 Industrial revolution

After this period a new age was dawning. It is the one known as the Industrial Revolution. It started around 1800 A.D. when steam engines were invented. Scientists and inventors proceeded to engineering machinery thus enabling the human species to carry out industry in a mechanical way.

The Earth provided us with raw materials used for generating energy and manufacturing products. The most important resources were coal and steel during those days. Previously, we were primarily dependent upon the supply of trees for fuel. One has got to keep in mind that in the past large parts of North America and Europe were almost completely covered with forests. This is a very important fact because plants turn carbon dioxide into oxygen the chemical element we need to breathe in order to stay alive. Without plants there would be no oxygen and without oxygen there would be no human life on Earth.

All living land animals exhale carbon dioxide. However, most of the quantity of carbon dioxide is released during combustion and production processes. Cars, airplanes, factories, the burning of forests in (sub) tropical areas to create farmland cause an excess of CO_2.

It took only half a century to construct a vast majority of factories. This rapid raise of industrialization led to a major change in society; urbanization. Industry generated a great number of jobs which attracted more and more people from the villages to the factories in the cities.

Conditions were poor in the beginning. Even women and children had to work.

Industrial power was positioned around a number of powerful capitalists. In the second half of the nineteenth century child labor was put to an end in the Western countries. Women were no longer required to work in the factories.

Nevertheless, the political power was a privilege to capital holders and land owners by means of the electoral census. Those who were able to pay sufficient taxes did have the right to vote. This system was replaced in most countries by universal suffrage around 1900.

In large parts of the world like South America, Africa and Asia there was hardly any industrialization between 1800 and 1900.

In South America and Asia industrialization started in the twentieth century.

Africa has since immemorial times to deal with the burden of drought and tribal wars. This has slowed down the development of progress enormously. Meanwhile multinationals try to take advantage of the natural resources in these continents. Examples include Nigeria which has a great amount of oil, and South Africa where large quantities of gold and diamonds are extracted from the mines.

Electricity and oil only started to play a meaningful role at the end of the nineteenth century.

1.8 The First World War

At the beginning of the twentieth century the political situation was unstable in Europe.

The assassination of the Austrian Crown Prince Franz Ferdinand was the fuse that caused the powder to explode. From 1914 to 1918 the First World War was raging across Europe although Germany and France were the main participants. This war was the result of a dispute over a piece of land between the two countries.

The Netherlands were neutral during the war. They were not involved in any fighting. But many countries throughout the world joined either of the countries. As a result almost 75% of the world population was involved in this war. Because there were hardly any tanks or aircraft devices in 1914 it was man to man battle. It was also called the trench war. Large areas on the battle zone consisted of trenches and barbed wire fences.

Ultimately, the war resulted in a victory for France bringing about more than 9 million casualties, soldiers as well as civilians.

Due to losing the First World War the German emperor fled to the Netherlands which held a neutral position. Despite losing the war the German population claimed victory over France. Therefore the German government was unwilling to pay retribution to the allied nations. Negotiations turned out to be poor for the Germans and as a result they still had many years to carry the burden of this "legacy" of the First World War.

Characteristics of this war were the trenches, barbed wire fences and fortifications. Poison gas, artillery, rifles and bayonets were the main weaponry for mass destruction. The front line was rather static. For many years the opponents besieged one another from the trenches.

So many people in Russia lost their lives due to this war. It led to the dethronement of the Tsar and the rise of communism under Lenin. Still, this was communism in the form of dictatorship because opponents were literally removed from society. Under Stalin this form of oppression was carried out in an even more brutal way.

The large number of casualties as well as the mutual distrust between the European countries also gave rise to the Great Depression at the end of the twenties and early thirties.

1.9 The Great Depression and the Second World War

The Thirties and the Great Depression

Industry was the creative force in the twentieth century although during the thirties people suffered from a huge economic crisis. This financial down fall started off in the United States of America.

During the First World War agriculture in the U.S produced food for the European countries because they were not able to grow sufficient food supplies themselves. When Europe recovered from the impact that the war had left behind the European nations became self-supporting again. Thence, the United States were faced with an overproduction of food. In addition, the American trades unions did not have much power and influence which prevented salary increase for employees. The domestic spending became less and less. In 1929 the bubble burst on the main stock exchange in the United States and the Great Depression was a fact. The impact was profound and the effect lasted for many years.

Thus, Adolf Hitler took the chance of inciting the German people in order to acquire their political support Hitler succeeded in realizing his fascist ideas by means of his warmonger proclaiming 'ein Drittes Reich', one large Germany from which all German citizens would profit.

The Second World War

In 1939 Germany attacked Poland starting the Second World War. The international community hardly responded to this act of violence. Germany continued to attack other countries as well and the ongoing German warfare made the rest of the world realize what was happening.

The Nazis believed the Aryan race (white skin and blond hair) to be superior to all other races. The Nazis carried the swastika as symbol of their ideology. Hitler blamed the Jews for all the economic problems and he managed to convince many people in Germany and abroad that the Jews had to be eradicated. This obnoxious point of view led to the found of concentration camps in which Jews were killed in gas chambers. Jewish people were also exposed to torture and medical experiments. The Jews were hunted down from the biggest city to the smallest hamlet all over Western Europe.

The Nazis declared themselves to be the new rulers of Europe and enforced their claims to this cruel reign by means of the above mentioned belligerent deeds. They were held responsible for the genocide of 6 million Jews. The extreme pain and grief inflicted on the victims and their surviving relatives is enormous and never to be forgotten. Thus, the Jewish people will always remain a nation with a sad history. It is noteworthy that this nation despite its limited size (currently estimating 15 million people worldwide) often played an important role in history. It still does today.

For decades the Jewish people have made the lives of many Palestinian people miserable which is a sad thing to see while the Jews themselves have been exposed to injustice for a long period of time (a clear example is the wall on the West Bank).

Hitler engineered the occupation of most of Western Europe. His winning streak stopped at the Strait of Dover. England did not surrender.

His 'blitzkrieg' stopped in the East in Russia by the onset of winter and long supply lines. The frontline was simply too long.

Japan used the ongoing international confusion to attack China and the United States. The infamous sudden attack on Pearl Harbor resulted in the loss of many Americans lives. The world was in chaos. It was the beginning of the Second World War.

The entire world was involved in this conflict. There was war on many front lines and battle zones. The Pacific Ocean, Indian Ocean, the Atlantic Ocean, in Asia, Europe and Africa were regions at state of war.

The abuse of power and violence was profound and devastating. Many civilians were maltreated and killed in concentration camps or lost their lives and their homes by bombardments on cities. Many people were buried in mass graves all over the world.

In order to top Japan from making war the Americans dropped atomic bombs on Hiroshima and Nagasaki in August 1945 causing hundreds of thousands of casualties most of them were civilians. A few days later Japan surrendered.

In the first half of 1945 Germans were pushed back and finally defeated on their own territory. Hitler and his newly wedded bride committed suicide a few days before the German capitulation.

The international community decided that such a horrifying war should never happen again. The Second World War ended in 1945. Germany and Japan were defeated. Europe was liberated and the Allied forces (Americans, Canadians, Russians, Poland, English, French, etc) returned to their home countries. This war had truly been against oppressors and it focused on freedom for everyone.

Only a small number of the ones responsible for starting the Second World War were brought before trial and got sentenced in a court of law. Many Germans, war criminals and collaborators fled in many directions. Many of them were caught in South American countries such as Uruguay and Argentina. However some of them are still living unnoticed in Germany today.

Some German scientists took up scientific research for Russia while others were employed by the U.S. A number of these researchers turned out to be the founders of space travel in the decades to come.

The most celebrated rocket scientist is Werner von Braun who designed the German V2 rocket during the Second World War. In the Sixties he designed the Saturn rocket which the Americans used to travel to the moon in 1969.

Switzerland was one of the few countries which maintained its neutral situation during the Second World War. This country is well known for its banking secrecy. For many years it has been a safe haven for criminals and laundering lawbreakers.

Some Jews who died in the German concentration camps had huge bank accounts in Switzerland. It was decades later when some of the money was paid back to relatives.

The reconstruction

Soon after the Second World War ended rebuilding activities started throughout the world. A new infrastructure was developed to restart the economy so that people could earn a living once again.

The Marshall Plan was of great importance to Western Europe. The United States supported European countries which were trying to stand on their own feet, financially. This financial aid was also set up in order to keep Western Europe away from the communist influences.

After the Second World War Stalin instantly established communist regimes in the countries that had been liberated by the Soviet Union army.

The world was to be divided into opposing camps; a capitalist section in the west and a communist area in the east.

Opposition to the communist system was brutally suppressed. Oppression particularly occurred in the Soviet Union and China. Each communist country had some sort of secret police and intelligence service. These institutions were organized with regard to detecting counterspies. However many civilians were also victims of these services. During the fifties the USA were notoriously known for hunting down alleged communists.

Key players in the capitalist system were North and South America and Europe.

Key participants in communism were China, Russia and Cuba. The rest of the world was the stage for fights between communist and capitalist imperialism.

The most distinctive symbol of these conflicts was the wall in Berlin between the eastern and western sections. The wall represented the ultimate segregation between two different ideologies i.e. capitalism and communism. This separation was also known as the Iron Curtain and it ran from Germany to the Balkans.

The forces of NATO[2] led by the United States and founded in 1949 were

2 NATO = North Atlantic Treaty Organization, the military umbrella organization of Western allies

stationed on the western side of this dividing line. The Warsaw Pact founded in 1955 under the leadership of the Soviet Union had posted its troops on the eastern zone[3]. This political collision precipitated an arms race which prompted the deployment of large numbers of nuclear missiles on either side several decades later. Fear reined the face of the world.

The independence of India

Another event less often mentioned but from a historical point of view as interesting as the developments in America and Russia is the struggle for independence in India.

At the end of WW2 India was still a colony of The Commonwealth. After a non-violent uprising the British rulers divided India into two main parts; Hindu India and Islamic Pakistan. The most prominent political leader of non-violent resistance is Mahatma Gandhi.

He was raised by the principle 'do not hurt any living beings'. Thus he became vegetarian.

A well-known statement of Gandhi is: 'The world has enough for everyone's need but not for everyone's greed'[4].

In 2008 India had nearly 1.2 billion inhabitants and it is by number the second largest country in the world. Only the People's Republic of China has more inhabitants, nearly 1.4 billion.

1.10 The fifties

War in Korea

The most crucial conflict in the fifties was the war between North and South Korea. After the Second World War North Korea was occupied by the Soviet Union whereas South Korea came under the direct influence of the United States. Again it was capitalism versus communism. When the two superpowers retreated from their satellite states the leader of North Korea decided to launch an attack on South Korea. A military conflict was the result. Political, financial and military interests had to be defended. The Western Allies supported South Korea while North Korea was supported

3 Soviet Union = Union of Soviet Socialist Republics (USSR) tightly led by the government in Moscow.

4 From the Dutch Wikipedia-internet site about Mahatma Gandhi.

by the Chinese who were afraid that war might reach their territory.

North and South alternately gained territory and eventually the war stopped. A final peace treaty has never been signed. Both armies are still standing opposite one another ever since.

The effects of this war were evident[5]:

- The United States increased their defense budget which started an arms race;
- NATO was expanded and the United States built military bases in Western Europe;
- A general feeling of fear imbued the West.

The occupation of Tibet

The People's Liberation Army of the Republic of China occupied Tibet In 1950. It is still not clear what the real reason for this act of violence might have been. The Chinese administration still claims that Tibet originally belonged to China. Like many other countries Tibet has a long history. Throughout the past Tibet has been occupied by various rulers such as Mongolia, China and other nations.

In 1911 Tibet became a sovereign state lead by the Dalai Lama, the spiritual leader of this country. One might question whether it is reasonable to state that an occupied nation should be overruled forever. It might be wiser and fairer to respect people's desire for freedom and have friendly relations with them.

Something else plays a role regarding this issue. Both the Warsaw Pact countries as well as he NATO nations surrounded themselves by buffer states shielding off potential opponents in the vicinity. The same strategic approach was carried out with regard to potential opponents in other parts of the world. The Chinese occupation prevented Tibet from becoming such a buffer country for the United States or England.

The Netherlands and Indonesia

In the fifties minor armed conflicts occurred, the Dutch police action in Indonesia being one of them. This hostile event marked the very end of the Dutch rule over Indonesia which was until then a colony of the Netherlands.

---

5 From the Dutch Wikipedia-internet site about the Cold War.

Furthermore, the fifties are characterized as the rebuilding years. Economies emerged and the world, as we know it now, was being molded. Many children were born which signified the baby boom of the fifties. Moreover, entire industries and cities were rebuilt. Last but not least the fifties will be known for a quiet period in world history.

1.11 The sixties and seventies

Racism and space travel

Racism and the struggle for power were important issues in the United States at that time. A lot of black citizens and supporters of communism were maltreated.

Everyone wanted a place in post war society. Everyone wanted a share of the new wealth but not everyone wanted to share that wealth.

This situation went on until the end of the twentieth century. By the turn of the century racism diminished gradually, but only because the United States had to deal with other issues such as Islamic terrorism and the economy going downhill.

Space travel was another topic related to the struggle for power. The Russians were the first to launch a satellite (Sputnik) and to send a human being into space. Yuri Gagarin was the first man to orbit around the Earth. The Americans nevertheless, were the first to land a man on the moon. Neil Armstrong was the first human being to set foot on the moon in 1969.

Fashion, music and lifestyle also became major topics in the sixties. Elvis Presley, the King of Rock and Roll, is still one of the most famous people of this era. He was well known for his modern way of dancing and extravagant clothing.

President John F. Kennedy

The most celebrated American president is John F. Kennedy. The most critical moment which he faced as president of the United States occurred during the missile crisis in Cuba in 1962. This crisis almost led to an atomic war.

The political opponent of John F. Kennedy was Nikita Chroetsjov leader of the Soviet Union. Russian nuclear missiles had been installed in Cuba and Kennedy requested Chroetsjov to take them away under threat of an armed intervention. Chroetsjov was intimidated by the warmonger Kennedy and the Russian leader pulled back the missiles from Cuba.

In 1963, soon after the Cuba crisis, Kennedy was assassinated during a public drive through Dallas. There is still a great amount of discussion going on about who actually killed Kennedy

Many official documents on theories and conspiracies of the killing of Kennedy will be released in 2029.

Robert Kennedy became senator and like his older brother also had presidential aspirations. He was murdered five years later.

At the end of the sixties the illustrious black reverend Martin Luther King was also murdered five years after the killing of John F. Kennedy. It is not clear who killed MLK and why he had to die but it shows us how depraved the American society was.

How is it possible for a country to murder its best people, the ones that tried to make a positive contribution to society?

The war in Vietnam

The Korean War was a small exercise for one of the major conflicts in history i.e. the war in Vietnam. The clash between North and South Vietnam started around 1965. South Vietnam was supported by the United States and North Vietnam had the protection of the Soviet Union. Again this is a chasm between capitalism and communism and it is about defending national interests.

The deployment of weapons was enormous in comparison with the war in Korea. Notorious were the bombings by B-52 bombers which were called 'carpet bombings'. Jets used to drop napalm on the front lines for the enemy to be burnt alive. Crop spraying planes were equipped with Agent Orange / Green / Red / etc. to defoliate the jungle so that the enemy became visible.

Sometimes a massacre took place such as the Mi Lai genocide executed by the Americans who were frustrated because of the invisible Vietcong (VC). This Vietnamese resistance army had built long corridor systems. By means of this tunneling device they were able to appear unexpectedly and disappear again just as quickly. The VC was actually a guerrilla army. Its soldiers did not wear uniforms. Therefore, it was difficult for the enemy to spot them. The VC was not to be distinguished from the Vietnamese citizens.

The average age of the American soldier in Vietnam was 19 years (in the

Second World War the average age of soldiers was approximately 27 years). Many of these young men gradually became insane during their tours of duty. Many of them were doing their service and had to go to Vietnam while during the Second World War most military men were recruits who volunteered. Unfortunately, the American people did not want to listen to their stories when they returned home. This was due to the increasing opposition against the Vietnam War at the end of the sixties and early seventies in the United States. The number of American casualties of what seemed to be an everlasting war went up rapidly.

The main problem about the Vietnam War was the fact that politicians imposed their will upon the army generals. This gave the VC a chance to repeatedly restore its front lines and supply chains. The Ho-Chi-Minh Trail was considered to be a lifeline for the VC troops. It was maintained that way throughout the war despite frequent American attempts to destroy it.

The flower power movement emerged and peace activists had many reasons for marching for peace and protesting against war because the Vietnam War was exposed on national television every day. Therefore, it was not a distant conflict but one any American citizen was exposed to in everyday life.

The number of victims coming back from Vietnam in body bags increased every day while the end of war was not at hand. More and more parents wondered what their sons were dying for.

Under great international pressure and by protests from the American people the Vietnam War finally ended in 1974. It ended with the withdrawal of the Americans from Saigon. Some 60.000 American soldiers were killed. Many of the ones who returned from that war still suffer from mental disorders and physical diseases as a result of violence.

The negative attitude of the American population at the end of the war against the returning veterans did not contribute to the process of rehabilitation either. In an extreme form this aversion is best shown in movies such as 'Rambo' and 'The Deer Hunter'.

Foreign workers in the Netherlands

During the seventies a number of people from Surinam and the Netherlands Antilles came to the Netherlands to earn money. Still, the Dutch needed more people to keep the economy going which was now running at full speed.

So the Dutch government decided to recruit "temporary" workers from other countries mainly Morocco and Turkey (as well as Spain, Italy and Portugal). Temporarily employed implies that these workers should go back to their own country after some time. However, a large number of the foreign workers decided to settle in Holland permanently reuniting their families in the Netherlands.

Things didn't work out that well for a lot of the new immigrants. On the one hand the immigrants had jobs no longer wanted by the Dutch. On the other hand they were the ones who were the first to get fired in times of economic depression. Often they could not find jobs again due to lack of speaking Dutch properly and lowly qualified training. This resulted in a rapid rise in unemployment among immigrants.

Moreover many people from large groups of immigrants could not leave their original culture behind which made integrating into Dutch society a difficult task to perform.

The problems for the immigrants became larger and larger every year. Unfortunately, the Dutch government responded inadequately to these problems.

Social services

On account of strong pressure from trade unions, a highly improved economic situation and especially a left wing government a large number of measures were taken to rally the position of the foreign workers. These measures included unemployment benefit, old age pension, disability remittance, medical schemes, and retirement allowances.

The economy repeatedly showed some downfalls but that did not seem to matter.

The natural environment seemed to be safe and sound. There had been a few small incidents but there were no large problems on the horizon.

In 1800 there were approximately one billion people in the world. In the early seventies the world population had increased up to approximately four billion people.

In the seventies the Netherlands had a population of thirteen million people. Left and right-wing governments were alternately in charge. This caused some kind of balance. The population grew slowly in the Netherlands and although this is a small country it caused no trouble. The family was the cornerstone of society. Women took care of the children and were in charge of the household. Men had a fulltime job. Every household had only one car.

1.12 The eighties

Wealth became a commonplace during the eighties. The flower power movement came to an end and a new music sounded through the eighties. Disco made its entry started off by John Travolta and the Bee Gees (Saturday Night Fever).

Elvis Presley got himself killed by an overdose of drugs and pills. People became aware that getting an overdose of anything is a bad thing.

The use of electronic gadgets and devices came into fashion. A large number of new commodities were invented. Most of these new products were luxury goods and made life easier and more attractive. The use of personal computers and cell phones took off.

Collapse of the Soviet Union

The Soviet Union seemed to go through a reverse development when Mikael Gorbachov entered the international political scene. His influence on Russian politics was signified by the scaling of the Berlin wall in 1989. The importance of his political ideas was also shown in 1991 when West Germany and East Germany were reunited. There was a significant crack in the communist bloc which led to the disintegration of the Soviet Union. Communism appeared to be an insufficient system.

It was the onset for a lot of changes. The power of the Russian mafia is significant and notorious all over the world. Actually, Russia was bankrupt. Due to huge Siberian oil and gas resources Russia still maintains an important player in the world economy. The western countries and China still want to do business with Russia because the oil reserves in the Middle East are decreasing fast.

The Russian population got divided into a large group of poor people and a small group of very wealthy people. Even most academics in Russia can barely earn a living. Despite two jobs and two salaries they can only afford a simple apartment for a dwelling.

The eighties ended with a sense of victory for the West and capitalism. Countries like China and India are still on the list as developing countries.

1.13 The nineties

Environmental awareness

The awareness that mankind is depending upon the natural environment of the Earth begins to grow in the nineties. Man also realizes that he cannot go on exploiting the planet at any cost. The list of endangered species becomes longer and longer. The ancient forests and tropical jungle are rapidly decreasing as a result of massive logging.

It becomes clear that the ozone layer is affected and that the North and South Pole are melting. Eternal mountains of ice and glaciers such as the Kilimanjaro in Tanzania begin to disappear. Scientists and organizations such as the World Wildlife Fund and Greenpeace are calling for action.

The United States economy still stands thanks to all sorts of protectionist measures. Still, the U.S. does not want to invest in environmental measures. Many factories are outdated and investments to innovate these production plants are low because the government protects them. Cleaner factories mean investing hugely. However, this would cut off the profit, hence the high share prices. The economy must keep on running at full speed from preventing the U.S. from losing its leading position.

Europe does not react to this financial downfall. The European countries know that the competition never sleeps and the continent is looking for cheaper ways to produce commodities.

Low-wage countries

Factories gradually move their production lines into low-wage countries as well as other labor intensive production activities. A country like the Netherlands changes from a production oriented society into a service oriented community.

Many elementary jobs disappear from the job market. Therefore, companies have a greater need for employees who are highly qualified and who know how to communicate properly about intricate matters.

Civil War in Yugoslavia

In the second half of the nineties a civil war in the former Yugoslavia took place. People from different ethnic backgrounds got into an armed conflict. The genocide committed by the Serbs against the Muslim population is the crucial issue in this war. The reasons for this war date back to 600 years ago when the Muslims enforced their place amongst the Christians.

This civil war exposed scenes which reminded the world of the concentration camps from the Second World War. The Netherlands were breaking news in a negative way because they failed to carry out the peace mission. The Dutch blue helmets could not prevent the Muslim population to be handed to the Serbs. Finally the Serbs murdered approximately 8000 Muslim men. The West intervened far too late and many of the war criminals have not been arrested yet.

Genocide in Rwanda

A few years later something similar happened in Rwanda where Hutus attacked Tutsis. These assaults resulted in approximately one million casualties. Again the West stood by, watching without intervening during the horrors of a civil war. The West did not act upon the Rwanda disaster because they probably had no interest in this area.

The First war against Iraq

Iraq raiding Kuwait provided the western allies with good reasons for launching an attack on Iraq. The NATO forces executed operation Desert Storm in 1991. The Iraqi army was defeated but the regime of Saddam Hussein maintained in power.

Generally speaking the nineties may be considered as a period in which the West celebrates its victory over communism. Countries like China and India are no more than sleeping giants.

1.14 The 21st-century

Terrorism

The leading country of capitalism is the United States and the American people went into a nationwide shock at the beginning of the 21st century. The once so solid foundation has become quicksand and no one really believes that the United States are able to restore the damage done. On 9 September 2001 Taliban terrorists attacked the United States by hijacking large passenger airplanes. They took over control and flew the airplanes into the Twin Towers and the Pentagon. Terrorism seems to be everywhere but invisible to the American eyes. America posed as the terminator of terrorism anywhere in the world. It is a good reason for attacking alleged terrorists all over the world.

First of all, the United States attacked Afghanistan. This country was believed to be the resort for Taliban warriors. In the eighties the Russians occupied Afghanistan but without succeeding to rule. The Americans seemed to be more successful in driving the Taliban from Afghanistan.

In 2003 the Americans designated Iraq as an ally of the Taliban. Iraq was also accused of hiding dangerous nuclear, biological and chemical weapons. Therefore, it was considered to be a menace to the state of Israel. An ultimatum was offered to Iraq. Iraq did not meet the terms imposed by the Western allies. Thus, the United States and Great Britain accused Iraq of lack of cooperation. They decided to invade Iraq and tried to arrest its dictator Saddam Hussein. He escaped at first but was finally captured and killed in 2006.

The economy of the United States

Going to war costs a lot of money and it causes enormous debts which can hardly be repaid. Large parts of the infrastructure in the U.S. are poorly maintained. Health care and decent social benefits are not parts of the social security system and are virtually nonexistent.

The United States are actually kept going by the rest of the world which tolerates these debts. These countries are afraid that their own economies will suffer too much from a collapse of the United States. The U.S. has a current debt (early 2008) of 2 billion dollars per day on interest. Despite President George Bush decision to give 100 billion dollars to citizens as stimulation for the American economy the financial situation nearly comes to a standstill as a result of the mortgage crisis. For a long time American citizens have been allowed to have a mortgage on condition of earning two salaries. This banking principle raised housing prices, pushing them sky high.

A great amount of risky structures were allowed including two mortgages on one house, pensions used as a mortgage and people without steady jobs having mortgages.

All these financial constructions were used in order to ensure that people would get mortgages so that money kept flowing. However, people who got ill and had bad luck in other perspectives might no longer be able to pay the mortgage.

There is a lot of work for temporary laborers in the United States. These 'temporary workers' usually pay more interest over a mortgage than workers with a permanent job and a high income. With regard to paying the bills a severe problem occurred. The problem became even worse when it became clear that mortgages for temporary workers were traded between banks and other financial companies all over the world. In short, a complex situation surfaced showing results which could not be predicted properly. Trading mortgages seemed a smart solution made up by clever people but eventually it did not turn out to be a remunerating banking activity. Risk management seems to be a line of duty they have never heard of.

The financial collapse of the United States will have a major impact on the entire world because of mutual investments. Nevertheless, the economic game still goes on and the budgetary focus is once again on new economic centers for example China, India and Europe.

The European Union (EU)

Europe is not a very large continent but the cultural differences are much greater than in the different states of the United States. Besides, one half of Europe is rich and the other half is poor.

In 1960 a number of rich European countries formed the ECC. The founding of this organization caused even more prosperity in the rich countries.

In 1993 the countries taking part in the ECC changed their organization into the European Union including some new participants. It set off major changes which started in 2000 when more and more poor countries were admitted. The social security arrangements of the workers in the wealthy part of Europe have been rapidly degraded in order to keep companies profitable. Expensive workers are being replaced by workers from the new countries. The indigent part of Europe is benefiting from this new situation but only because it is a cheap trick. The rich companies will do their utmost best to keep it that way. Therefore, the integration of European countries is a difficult process.

There is still a lot of debate going on about the participation of Turkey to the EU. Turkey is infamous for the supposedly lack of human rights. Nonetheless, do we really care about this matter? While the Turks were bombing the Kurds in the north of their country, many Tourists from all over the world were on the West Coast on the beach enjoying their cheap vacation.

Economical growth in China

The economical growth of the new European countries is limited partly by the rise of even cheaper emerging superpowers i.e. India and China. China does not care about human rights. A good example of how the Chinese government authorities abuse people is the mistreatment of prisoners. Whenever a shortage of donor organs in hospitals arises or whenever hospitals need large amounts of money organs are being surgically removed from prisoners.

The Chinese government has supreme power and if necessary the administration will use that force to suppress any opposition.

In recent years the military build-up of the country has progressed enormously but strangely enough the international community did not seem to be alarmed by this expanding military force. The rest of the world is more concerned about the economic potential along with the growth of market areas. Every prosperous country wants to do business with China despite the fact that China occupied Tibet and swallowed up Hong Kong and also wants to take over Taiwan.

The development of India

India is another important developing country. The infrastructure is minimal. Floods and earthquakes regularly ravage the country. The government is hardly in a position to take measures themselves it seems but it has sufficient financial resources to build up a weapons arsenal and to fight with neighboring Pakistan every now and then.

Today, India has a thriving economy with high growth rates which are almost equal to those of China. A relatively small group has become very rich in India. Yet, Western society is still sending money to poor parts of India as if India suffers from lack of financial resources.

Furthermore, some Dutch people are losing their jobs to laborers in India because wages in India are lower. But that is something we must appreciate also, it seems.

We hardly look at the caste system in India which is the reason for the suppression of a large section of the population. Officially this system does not exist anymore but in fact it is still very much alive. The caste system has been derived from the Hindu culture and this structure was to illustrate how the Hindu society should function. Each caste was equally important just as any man is important. Instead it was introduced as a class system.

The lowest class is caste-free. Nowadays, this class includes approximately 160 million people. They do the dirty work for which little or no training is needed. Even cattle are treated better. One cannot move up to a higher caste; the caste is a lifetime inheritance. Children from lower caste families are in the same rank their parents are in and cannot transfer to another caste.

The vulnerable balance

Meanwhile, the Earth seems increasingly out of balance. Today, there are more than 6 billion people on the Earth and the human species is rapidly growing up to a number of 7 billion people.

Fossil fuels still form the basis for our economies. More and more countries increase their productivity which causes many problems. The number of storms and hurricanes is increasing. The average temperature is rising. Sudden weather changes are increasingly becoming part of the meteorological system. They appear at times and in places where they previously did not occur. The Earth is warming up. Air pollution and water contamination are already major obstacles. It is increasingly difficult to get healthy, non-manipulated food. Prosperity seems to be a huge success but does this also apply to welfare? Critics keep asking this question over and over again.

1.15 Lessons from the past

What can we learn from world history? What can we say about the development of humanity? What are the characteristics of human behavior throughout the years?

Wars

The wars in Afghanistan and Iraq remind us of the Crusades in the middle Ages. Jerusalem is the most important and holiest city to Christians. Therefore, the initial aim of crusading was the liberation of Jerusalem from what Christians believed to be heretics. Mainly adventurers went to the Holy Land. Their main purpose was fame and fortune. They wanted to be heroes.

During the times of the Crusades looting and massacres were common ways of warfare. Because of the cruel and brutal war crimes executed by the Christian armies Crusades rapidly lost their noble goal.

Afghanistan being one of the most corrupt countries in the world was attacked because of the Taliban. The Western allied nations strived for a quick deletion of Afghan corruptness.

But is that attempt the real reason for invading Afghanistan or is the true justification for the attack the defense of the American economic profits prompted by an important project with respect to a gas pipeline?

The allies attacked Iraq because they thought to reveal evidence of nuclear, biological and chemical weapons. According to the United States, these weapons of mass destruction were a threat to world peace. However, no NBC weapons were discovered. The suggested relationship between Iraq and the Taliban could not be proved either.

Are the huge oil reserves the real reason for the war against Iraq? The United States themselves have difficulty obtaining oil supplies because the American oil wells are drying up.

All these wars are typical for human history. It has always been a matter of interests and power. Waging wars is rarely a matter of freedom for people. Only the liberation of Europe during the Second World War is really about liberty of mankind.

Great inventions

The biggest change throughout history was above all the invention of the steam engine in the second half of the eighteenth century. It resulted in the industrial revolution and gave the development of the cities a boost. We went from being an agriculture-oriented society to an industry-oriented society.

A second great invention was definitely the computer which actually is a highly simplified version of the human brain. Today, a lot of machines are being controlled by microchips, some of which are plain, simple devices while others are extremely complex micro instruments. Without computers today's society would no longer be able to function properly. By means of computers we were able to transform our society into a service delivering community at the end of the twentieth century. Mankind invented many more important tools but nothing affected mankind to a greater extent than the invention of computers.

From this point of view the capitalist system did not bring a great deal of progress.

That is why you see that we are now at a bit of a dead end, and must come back from earlier inventions; think of Nuclear power, the Concorde, the SR-71 Blackbird and the Space Shuttle. These are all great inventions, and the best of their kind. However, these technological highlights are extremely expensive and they are certainly not without risk.

Sociological perspective

From a sociological approach things do not seem to be right. Apparently, a small group of people are still trying to suppress larger groups of people. Subsequently, people who are being oppressed rebel against their oppressors. This resistance results in groups of people complaining without realizing that they are causing the resistance themselves.

People who live a good and fortunate life do not go to war. Neither do they become terrorists or criminals because they have too much to lose. Do we realize what's at stake? Or do we believe that everyone gets what they deserve? Do we really think that people become terrorists or criminals for the sake of entertainment? Do we truly suppose that people are to blame for poverty? Aren't winners and losers just parts of the system? Is it an inner urge that people want to protect themselves from living in poverty even if that means that they have to lie and deceive? The situation in which people find themselves is the result of the system that we are living in. It is rarely the choice of people themselves. In a different system these people would probably find themselves in different situations.

Mankind should choose the system which allows everyone to be happy. Apparently the human race does not make that choice. Exercising power over people might make people act. People acting out of fear for losing their jobs will not be happy. Thus, exercising power over people is more important than people being happy. Still, this mechanism is an evolutionary result and it has been demonstrated throughout human history. Humanity has not made sufficient efforts to change this situation. Instead, mankind still walks beaten tracks. Status, money and power, often accumulated in the course of hundreds of years, are still not being shared equally. If someone succeeds in becoming a member of this rich and powerful group the person involved wants to enjoy this wealth as well without giving up what has just been acquired. Outsiders should not enjoy wealth unless they have achieved similar goals. This results in a situation in which the happy few have billions in their bank accounts whereas the majority of the population lives in poverty. The money of the rich and famous is not useless after all because it is used for securities trading, making even more money.

These people are not producing goods or food they are only making money with money to have more money in the end. For what reason? If the stock exchange falls deeply, huge amounts of money will be transferred to banks and financial institutions in order to make their shares rise again. Who profits most of this?

Development and change

At least, we should learn one important positive aspect of the history of the world; change is inevitable. Or rather, development is inevitable. We have to realize that development also implies change. Something that develops will not remain the same. Changes may have either positive or negative effects. Something that evolves needs space and freedom. Limitations to growth or change will not be successful. We may try to guide it in the right direction although offering too little space will produce resistance. That will be harmful to someone or something. This dogma applies to both humans and the Earth.

On the one hand, the aims of the communist system and the goals of religious states will not be effective in the long term. On the other hand, the capitalist way including abundant decadency and greedy individualism will neither be sufficient in the future.

It is a fact of life that things will not stay the way they are now. However, if mankind does not intervene, changes are likely to be out of control. Instead, changes will be determined by the out of balance factors. The outcome might be disadvantages for a large number of the people, if not the majority of mankind. That's why it is better to change the current situation into a balanced condition to the benefit of the entire world population. Until now, we have never succeeded in doing something like this and it has always been the law of the strongest, as in nature.

I for one believe this is the chance for humanity to show that we have indeed learned from the past and have gone further on the road of human development. For once, this is a chance to make a conscious choice to change the world instead of just letting things happen.

Spirituality

The fact that man has believed in a 'God' and/or life after death from the beginning of time is another positive conclusion to go on with.

Man has always had the feeling that there is more between heaven and earth than we know. It is unfortunate that religion is so closely related to wars and cruelties which have taken place over the years. By giving presents and money to charities or by prayers, we are still hoping to get forgiveness for our sins. After repentance we happily go on with cheating and deceiving and suppressing people. I do not think that anyone who practices faith in this way knows what he is doing. The purpose of most religions is to let the love for our fellow men prevail, so that people will become happier. There must clearly be some kind of positive development.

Consistency is created by freedom. The exercise of power is no basis for the long term. We have seen this time and time again. A healthy loving relationship is characterized by giving and not by taking. The best way to create lasting relationships is by setting people free. They will appreciate this and they will always come back. It will not work if people feel restricted by rules and laws and continuously must meet various obligations.

Most important conclusions

Summarizing chapter 1 shows us a number of central themes to identify:

1. the behavior of humans has hardly changed in the course of history;
2. mankind mainly acts because of his own interest;
3. humans seek status (who are you, what do you own, what do you do for a living) and power (how many people does one control or strongly influence);
4. money and power are causing violence and crime;
5. status and power are gained even at the expense of the 'happiness' of fellow human beings and a society in 'classes' is created;
6. financial profit is more important than principles and welfare;
7. the current society is focused on growth, not on balance;
8. the government policy of the current society is primarily focused on the short term, it lacks a vision for the long term;
9. maintaining the current system is more important than the happiness of all people;
10. change is (eventually) inevitable;
11. change means most of the time progress;
12. people have always believed in a 'God' and a life after death.

Recapitulating, we might state that there is a basis from which we can work but we are still lacking the proper completion. It is important to conclude that most people:

a would like to be happy;
b believe in love;
c like progress;
d believe in a 'God' and/or life after death.

The above mentioned outline will be used later in this book to describe the transition to the spiritual society. First, we will look at other aspects which also point in the direction of the new society.

Chapter 2

Bottlenecks in today's society

2.1 Introduction

I described the situation from past to present in the previous chapter. This chapter examines what is wrong in the current world society. A lot of these problems are results from the past. Upon looking on what happened in the world I keep asking myself why our society is not a logical and fair system. Why do people hurt one another? Why do people not tell the truth? And why do most people still claim that they are happy? Is there another way? Science is one of the most important parts of the foundation of our society and yet this is rarely the basis for important decisions. Why not? Why do some people get paid a lot of money for doing jobs one does not need training for and which is completely useless after all? And why do many people who perform important and productive work get paid very little?

Money is a fact of life in our society. How can it be that we create and destroy money whenever we want to? These questions made me reflect on life. I will answer many of these questions in this chapter. It's impossible to answer all the questions about the world we live in but hopefully some things will become clear.

What I want to describe is just the tip of the iceberg. If I wanted to be exhaustive I would write many books concerning these questions. I am only writing about things we perceive but there are many bad situations most people including myself are not aware of. However, it is not my intention to be exhaustive of all wrongdoing but rather to clarify the positive alternatives.

Identifying the bottlenecks is only necessary to illustrate that we are getting further away from what most people really want.

I believe that we are now on the brink of world history from where we still have the possibility of giving a positive twist to the existence of mankind on Earth.

In Chapter 3 I will elaborate on what will happen if we do not act upon this right now.

The following paragraphs will depict many Dutch examples. The Netherlands is a characteristically rich and developed western country with a rich history. It is also part of the European Union (EU). The Netherlands has become well known in the course of history because of its role in shipping, trade, water management and innovation. The Netherlands also play an important role as an international partner for foreign countries.

It is a democracy with a multicultural population. The Netherlands is also one of the most densely populated countries in the world. The vast majority of people in the Netherlands have a good life. But is good well enough? Is such a life good enough for anyone in the world? To which extent is this country an example to other countries? Or do we need something else?

2.2 Norms and values

Norms and values shape our society, our living environment and determine how we may develop ourselves and be free, happy, and so on.
Values are aspects (motives, ideals) in our society which we are striving for. Included are justice, security, respect, happiness, freedom, love, etc. Although these values are part of the foundation of our society they are hardly described or recorded. Sometimes these values are being referred to as universal values. Each individual or group chooses its own specific values and adapts these values to their specific use. The consequences are obvious. If principles are interpreted differently the realization of them is also different.

Anyway, we are trying to achieve social values by means of norms. First of all, norms are laws and rules which are part of our constitutional system. People are supposed to be law abiding citizens. Laws are written down in law books or statute codes in order to avoid misinterpretations of the law. Therefore, there can be no misunderstanding about the meaning of the law. Generally speaking, violation of the law is followed by a penalty or punishment. This is common knowledge. Beside legal knowledge of the law society maintains social standards which are determined by the social role of an individual person and the expectations of the community. A social role is a socially defined pattern of behavior exhibited in particular situations and specific groups of people. Examples of social roles are the father and mother role model, single persons, traffic participants, Muslims, Christians, poor or rich people, boys, girls and so on. Based upon the description of the social role an individual draws a mental picture of expected behavior accordingly to a particular social role. The ultimate outcome depends on the knowledge and background of that individual in relation to the role.

Moreover new unwritten rules will appear within a group telling group member's how they should behave. These are social standards. Examples of social standards are keeping quiet in a library or church although this rule has not been written down. Another social standard is that the native population should be tolerant towards immigrants. Greeting neighbors is one more social standard. Different groups maintain different standards which might lead to conflicts.

Honesty

As a child I was taught to be honest and not to tell lies. Most people are honest and tell the truth most of the time. We also read newspapers and watch the news on television. After many years one might come to the conclusion that the world is not an honest place at all. Examples of dishonesty are:

- For the first time in history more relations break down than new ones are being registered. Many people who break up their relationships have not been honest to one another. At least 50 percent of married people who do not divorce have a bad marriage and stay together for a long time only to break up eventually.
- A large number of people commit economic crimes by obtaining unjust benefits.
- Managers line their pockets at the expense of their employees using prescience and inside information.
- Many companies employ illegal immigrants.
- Cronyism unveiled in many Dutch industries seems to be norm.
- Taxes are evaded.
- People lie and cheat in order to prosper at the cost of others.

The above mentioned examples show abuse without people getting injured or killed. But what about the allegations against Iraq expressed by the United States? Rumor had it that Iraq secretly had control of chemical and nuclear weapons. Is it righteous that this assumption was the justification for armed intervention? The assumption caused a war in which a lot of people were killed. Nevertheless, no nuclear or biological weapons were detected. Even the former president of the United States George W. Bush had to admit that the suspicion was false. Yet, he did not step down at that time. From that moment on his credibility was waning. No wonder he was well known under the motto 'if violence does not help, use even more violence!'.

Even worse, Chief Executive Officers who made huge mistakes retire with a big bag of money whereas employees will not be sent away with large sums of money when they are fired. If they are lucky, they will get just enough money to find another job within a few months.

Laws and regulations

If rules or laws are violated the values that stand above them are being violated as well.

For example, dashing across on a red traffic light is violation of the law. Such a risky traffic offense jeopardizes other road users. Security is the key value.

Danger is hidden in the relationship between norms and values, especially if we do not see it! A good example is scam money. No tax is paid for scam money. If the roads are getting into a bad condition society asks the government to pay more money for the construction of high quality roads. Due to tax violations there is less money available to keep the infrastructure of our society in a good shape. The central value is that everyone should contribute to society.

One of the consequences of the growing population of a country like the Netherlands is that we have more and more rules and laws. This increasing regulation only tries to ensure that all these people continue behaving in a somewhat tolerant and orderly manner. However, is this still what we consider to be freedom? All these rules force people to move around in a predefined way.

I will give an example of such conditioned behavior. Step out of your front door. Stop! Are you dressed according to the rules? Are you not naked? Being a man or a woman, do you wear proper clothes matching your gender? Is your face not covered too much (the police must be able to recognize you on video)? Do you have your identification papers on you? OK, now you are allowed to go out on the street. Walk on the footpath on the right side. Now, you have to follow a whole range of traffic rules. Being out in the street you might be robbed by someone who threatens you with a knife demanding your money. You are not allowed to knock this person down unless he hits or stabs you first. Yet, it is hardly possible to take away a knife from a robber without using violence. So you give him your money. The mugger runs off and you will go to a police station in order to report this crime. Do not think the police will come immediately to help you if you call them because they are much too busy doing other things. If you do not follow the right procedure for reporting a crime the police will not help you. This example shows that the bureaucracy in our society is endless.

It is easy to see that there are too many rules for a normal human being to know them all. Nonetheless, the number of police officers is not sufficient to keep situations under control. Apparently the system is not that good and it is apparent that order enforcement is symptom control. We should ask ourselves why we keep breaking rules only to find ourselves trying to analyze and abolish the roots of the problems afterwards. Here's a straightforward example. People will probably be tempted to drive faster than the maximum speed allows them to if they drive cars which are technically able to exceed the maximum speed limit. If cars technically generate only a limited amount of horsepower the maximum speed limit will not be violated because you will be at the next traffic light before driving faster than 35 miles per hour. Obviously, proper traffic education and awareness might contribute to safety as well.

Every day newspapers, television and radio shows tell us that people do not obey laws and rules neither do they care about values. Neglecting laws is usually a matter of money and power and ignoring rules is often about lack of respect for each other. Again, it comes down to the same old story. Still, we do not investigate the origins of the trouble. Instead we are only doing something about the symptoms. The next day we see similar news flashes and the day after tomorrow the same bad news is all over the media again.

However, honest people who report violations (they are so-called whistleblowers) are ruthlessly punished for reporting abuse and crimes. Most of them are dismissed from their jobs leaving them unemployed. Some of them have already deceased or suffer from serious diseases as a result from all the stress that telling the truth may have caused.

Social security system

A clear example of deteriorating norms and values is reflected in the social security and health care systems of the Netherlands. The Dutch government emphasizes the importance of norms and values in society. At first sight this approach seemed to be good. The social security and health care systems were changed a couple of years ago. But it soon became clear that norms and values were far less important than the financial situation of the government which was mainly supported at the expense of the achievements of the working class. The early-retirement and retirement schemes were abolished and employees are supposed to keep on working until their 65th birthday[1].

The process of individualization in our society is also incorporated in the tax-system in the Netherlands. Some years ago the Dutch social security system provided both partners with allowances as soon as the male partner became 65 years of age even if the female partner was younger than 65. Nowadays, retired couples will not get a pension until the female partner is also 65 years of age. This leads to a huge dip in revenue because the pension is still based upon two allowances from the start. Retired people have to fix this problem themselves.

With regard to the old age pension the Dutch government is working on an extra tax law for pensioners who are entitled to a company or additional pension scheme/arrangement.

On top of that, the Dutch government forbids the payment of financial arrangements before the age of 65 years. These payments were the same arrangements the administration encouraged at first!

To be short, a pension in 1990 (including the prospect of retirement before the age of 65) has been changed throughout the years. Nowadays, there is no guarantee for early retirement. Add to this tricky business all the promises made by banks and insurance companies claiming that all signed contracts will be profitable after 20 to 40 years. This will happen either due to the disappointing investments and because of the high costs insurers are charging. It is easy to imagine that the financial flexibility of employees evaporated at a large scale. Moreover, for many people the future does not look promising. People who retired during the last decade and made good financial arrangements do not have to worry about the future. Actually, they are contributing to the

1 And the mist has not yet been blown away about the obligation of people to go on working until they reach their 65-th birthday by mid-2008 and there are already plans being made by the Dutch politicians to make it compulsory for people to continue until their 67-th year of life.

economy. However, things are getting more difficult for the next generation. The dismissal of employees will be simplified too. The period and level of unemployment benefits have already been reduced. People who have fallen ill and depend on social welfare get less money than a few years ago. How do they have to pay for all the things they need in daily life?

Health care system

Keep in mind the revision of the health insurance system. We all depend on the health care sector because anyone who grows old will need physical or mental help sooner or later. Because of this the increase in prices in this sector are tremendous. Instead of controlling the costs, insurance contribution increases every year causing this sector to become less and less accessible for a vast majority of people in the future.

In early 2006 the health care system was adjusted and insurance contribution was revised. Of course it is the responsibility of the government and insurers to ensure that insurance installments are sufficient to cover all spending. To attract customers insurance companies lowered the insurance contribution to a great extent[2].

Early 2007 it appeared that insurance companies became short of money. It might seem logical for the insurance industry to solve these financial problems by themselves. Unfortunately, this is the wrong assumption because the deficits had to be paid by the insurance customers in that same year! The insurance companies did not accept lower profits and did not solve these problems, they had a commercial goal to bring in customers; unfortunately the individual customer paid the price. The list of such measures is endless.

Children and elderly

Attaching importance to norms and values implies educating people. The education of children is a convenient starting point for emphasizing the significance of norms and values. The fact that our society mainly shows an economic interest in people causes an increase in working hours which prompts

2 You are not automatically 100% insured. If you want to do that you need additional insurance. This is arranged in levels. You are always insured for the most important things.

less time for parents to educate their children[3]. It diminishes the chance for children to learn proper norms and values.

Many young people are hanging out in the street these days. We ignore the values they pick up on the street. If these youngsters stay at home they often surf on the Internet. Actually, that's worse than hanging out in the street because it is well known that bad things are exposed on the Internet.

It is common knowledge that wisdom and age go hand in hand. So listening to older people might teach people a thing or two. Elderly people used to be the corner stone's of values and standards in yesterday's society. How different this is for our modern society. Many workers who are older than 50 years get fired and are disposed off. They are said to be slow and rumor has it that they easily catch diseases. They produce too little and cost too much. Let elderly people perform simple, easy jobs such as delivering parcels or weeding parks. That is what they are good for! That is the way we take care of our elderly people nowadays. We do not learn from the lives of these people. Instead, we're throwing away their experience and wisdom.

2.3 The education of children

Labor Distribution

In this paragraph I would like to describe in detail the education of children. What does the situation regarding this subject look like?

In the first decades after the Second World War children in most West European countries, including the Netherlands were being looked after carefully. People were willing to trust society again and they thought about how to reconstruct their world. Many children were born in these days. This natural increase in population is called the post war baby boom.

In the late sixties and early seventies families decreased with respect to the number of children. Pressurized by scientific pedagogical theories and a growing prosperity fewer children were born due to parents who preferred to spend more time, money and energy on a small family. It was the onset for a declining birth rate.

---

3 The taking care of children in kindergartens is from a pedagogical point of view something other than raising children. The upbringing of a child is designed to ensure that he/she later can function as an independent adult in society. Various aspects can be identified: emotional education, aesthetic education, ethical education, intellectual education, physical education, personality formation, sexual education and social education.

Generally speaking, husbands had fulltime jobs and wives were in charge of taking care of the household and the children. An average working week consisted of an average of 40 hours of hard labor. Marriage was a holy institution. Therefore, breaking up a marriage was not done. Family life was the cornerstone of society. 'Sociability' was a real Dutch understanding. Breakfast and lunch for the entire family were central moments of the daily routine. It could easily be done that way because most people lived close to their work. Children were looked after by their mother's after school. They had tea together allowing children to tell about their school day. When fathers returned from work it was a general custom to have a family dinner. Most families only had one TV set. Therefore, watching the telly was collective leisure. Another pastime activity was playing card games or board games. Society was fairly homogeneous which contributed to harmony.

Production figures had to go up by force of economic circumstances. People were required to achieve more goals and "obtain" higher targets. The production rates of employees had to go up. This requirement for higher production figures related easily to the needs of the emancipation movement and feminist wave. These movements required that women and men were to be treated equally. Besides, emancipation and feminism are driving forces striving for equal chances and opportunities for women. Women also claimed the right to have jobs. This development of women's independence and self support led to changes in many family situations.

About 15 years ago it became possible to have a mortgage based upon the income of both partners in the Netherlands. This newly arisen funding construction was not possible until then. It signified that people could spend more money on buying a house. As a result, prices of houses increased and nowadays the only way for a couple to afford buying a house is a double income. In general, most Dutch men still work fulltime (36 to 40 hours) whereas Dutch women do have part-time jobs (averaging between 25 to 30 hours). Overall, the average amount of working time is between 61 and 70 hours per week. And in the Netherlands we would like women to work more hours per week. Compare the current situation to the circumstances of 30-40 years ago!

Most of the employed have to commute because they do not live close to the companies they are working for. This means a great amount of commuter traffic. Two or three hours a day is not out of the ordinary. This strongly reduces the amount of time spent on educating children.

We should rather address this way of taking care of children as 'child care' which many parents leave to others because of lack of time. It also implies that children need to be more self-supportive. Sociability in family life is hard to be found these days because parents are tired at the end of a working day. They would rather spend time on their own interests than sharing time with their children and family, because they want to live up to the modern standards set by society. Children are given money or have jobs on the side line. Do not think that child labor has been abolished. No way! Fourteen-year-old children may already get a 'small' job. It gives them something to do and earn pocket money. Company owners can still get cheap labor this way. So, everybody's happy, aren't they?

Behavior and violence

Juvenile problems are of a far more serious nature today than it used to be in the past. School students carrying weapons in school grounds have become a huge problem. The number of students leaving school before graduating has rocketed sky high and has become a bigger problem than ever before. Many youngsters are becoming looser and bolder. The rate of drug abusing adolescents intensifies. Violence among young people is taking place to a higher extent and taking on more severe forms. Youngsters participate in fights using knives and guns shooting pupils and teachers.

At the end of the twentieth century these events were exceptions but at the beginning of the 21st century it has become common reality. An additional problem is that many of these youngsters show an unscrupulous nature. The police, school leaders and politicians in the Netherlands have become extremely concerned about this tendency. This topic seems to be a part of the paragraph 'Crime' but is being discussed in this paragraph instead because I believe the violence amongst children definitely is caused by the decline of the proper upbringing of children. If we want children to behave respectfully and lovingly towards other people we ought to teach them the importance of these values. Neglecting children might lead them to bad habits such as watching TV or surfing on the Internet for too many hours. Thus, they may be exposed to violent movies and video games. Television and the Internet might become their real world. However, it is the primary responsibility of parents to ensure that their children will not be affected by the bad influences of aggression on television and the Internet. It might be possible to prove scientifically that watching violence on TV and the internet is related to the (violent and hostile) behavior of children although such evidence will not really be necessary. This relationship will be obvious by logical reasoning. Especially children from the weaker parts of the population and children who have not yet become psychologically stable will be influenced.

If these children successfully finish their school career, one may wonder whether they have actually learned a lot. The complaints about young adults who are not able to count or spell properly are becoming more frequent. Apparently when children have a paid job, it comes at the expense of the ultimate knowledge.

Quality time

'Quality time' is a concept which was first mentioned some 25 years ago. This concept originated in the United States and crossed the Atlantic a few years later. What does this notion mean? Putting this idea in simple words it means that parents are busy building a career. Therefore, they have less time for the upbringing of their children. As a consequence parents reserve a fixed number of hours per day at a fixed time for the children in their daily schedule.

Some American day care centers enable parents to observe their children by means of webcams to see if their offspring is doing alright.

Only a decade later, we learned that this concept was not working genuinely. Children cannot be programmed to be raised during fixed hours suitable to parents. Children need immediate attention and support if they come across situations they do not understand or if something goes wrong. It is also a psychologically established fact that punishment only works when imposed immediately after the act. The longer it takes before punishment is inflicted the less effect punishment will generate. People hardly relate to the act if it takes too long before being punished. If so, they will oppose against the person imposing the penalty. The same mechanism works for rewarding. In short, education will only have positive effects if parents take care of their children continually. People in the United States realized this too. In fact, woman would rather have part time jobs in order to spend quality time with their families. Thus, a counter-movement was founded striving for a more comfortable society. But we observed what happened in recent years. Outsourcing of employees and the emergence of low-wage economies such as China and India took place. The economy of the United States and other western countries suffered from high financial pressure. It demanded higher production volumes at lower costs. What happened to the counter movement? Nowadays, we do not hear a lot from the United States on the subject of education and women working as part time employees.

So we are still going in circles. Will society be able to raise children to become happier adults in the world of today than in the former millennium?

2.4 Interests

(Managing boards of) Companies do not always tell the truth. Information is held back or interpreted positively for allegedly protecting employees or the company itself. This is when we enter the field of interests. Most people have material and/or business interests regardless of what they are doing. The behavior of any individual is also determined by the (business) interests of other individuals. Sometimes the truth needs to be explained different from the real message. Yet, in this way we create a society which is so complex that we often find excuses to do or say something untruthful just because it is to our advantage.

Furthermore, we regularly shift beacons as a result of the ever-changing interests. Living such a life leads people astray and eventually people will realize that there is no turning back. This is a well known theme to be seen in many movies. The main character contemplates his personality and the impact of his actions. Along the way, the protagonist ends up reflecting his life. Obviously, this is not what happens in real life.

Self-interest, business interests and state interests

At the end of 2007 I watched a television program about people demonstrating against the use of windmills in the province of Utrecht in the Netherlands. Local demonstrators found windmills to be the cause ruining rural views! Yet, the Netherlands has been a country of windmills for many centuries. These devices were built in order to maintain the environment but this idea did not seem to enter the minds of the local demonstrators. It is most likely that they were demonstrating because they lived at the edge of the town. They might be worried about the value of their homes which may decrease because of the spoilt view. Do not think that this was the first demonstration against using windmills because there were many objections in the past for the same reasons.

In 2008 a research[4] revealed that the Dutch are less willing to contribute to a healthy environment than other Europeans. This rate goes up from approximately 25% of the Dutch to 50% of all Europeans! It is often noted that short term interest is more important than the interests of the future of our children.

An example of state interest is the number of soldiers suffering from cancer as a result of using uranium enriched ammunition in war. The number of these casualties has been augmented since 1991. Ammunition containing uranium is able to destroy the thickest armor of tanks. Therefore, these arms are most powerful weapons.

According to a report by the World Health Organization (WHO) it has not been proven that there is a clear link between the use of uranium and getting cancer. Nevertheless, in 2008 one of the authors of this report stated in a television program that the studies which supported the theory were skipped. The importance of the use of uranium ammunition is too big to eliminate.

A number of years ago a similar event happened concerning the HAWK-radar. The HAWK-radar was used to guide air defense missiles to low flying enemy aircraft. It was widely used by NATO countries. By now the HAWK-radar has been replaced by the Patriot anti-aircraft-system. In several countries people began to wonder about the remarkably high number of cancer patients among employees of the HAWK facilities. Was this high rate due to the radar-emission? Some of the sick soldiers sued the Dutch Defense department but investigations by TNO (an independent scientific research institute in the Netherlands) claimed that the possible impact of the radar was negligible. The sick military men lost the trial. Since then several veterans have died of cancer.

Still, the debate about the effect of radiation on the human body continues. Radiation is a complex issue just as the cause of cancer is. The link between radiation and cancer is extremely difficult to explain. Doubt is not an adequate reason for abolishing equipments that may cause cancer; business interests are all important. First of all, the assumption that radiation jeopardizes the human health must be made explicit by a research following scientific rules. As long as the number of casualties is not large, nothing can be done. Let us pray that none of our children will die from radiation! I may be exaggerating but

4 Nov. 2007: study by PR consultancy Porter Novelli of over 16,000 consumers of eighteen years of age in eight European countries.

it seems that we accept such risks more and more. These risks are the consequences of our level of prosperity.

The human interest

The worst aspect of our society is the slow but increasing decay of human interests.

People start to talk about 'economic interests', 'interests of national security' and 'business interests' to mention only a few examples. We use these words as an excuse for not taking the interests of individuals into account. They are crushed by 'the system' which has the highest priority. When the system generates an opposite priority people get crushed again.

Five months before the Olympic Games began in Beijing, the United States skipped China from the list of countries which violate human rights. I heard it on the news just after I had watched a Dutch television program about human rights in China. A representative of the United Nations explained that human rights in China are still not great (only to add that the human rights situation is much better than it has been for years).

Apparently, interests are also at stake here.

Why does all this happen? After a few more twists, nobody knows the answer. Besides, the majority believes that it does not matter. Most people do not seem to be concerned as long as their own needs are satisfied.

Many people think this is a ridiculous situation. Still, they let it happen because they feel powerless. Many of us think that managers, directors, shareholders do whatever they want to do. This seems to be true. The interests of these powerful people are so great that they have a lot to lose. It makes them dangerous to their fellow men because they will not hesitate to destroy other people's lives if they think that their interests are jeopardized in any way. They have money and power, they own expensive cars, large houses and women, ruling their self made empires. Do you really think they are willing to give up their wealth so easily when being told that they are running an unfair and evil business?

It's not the way these people think. If someone has the nerve to tell them, they will talk themselves out of it or consider the spokesman to be crazy. They believe that those who have wealth honestly earned it and that the rich are destined to be wealthy. They also believe that people who are poor only have themselves to blame and that a lesson is to be learned from living in poverty.

They try to justify and maintain the current situation instead of trying to change it. Some examples in this context are:

- In 2007 various countries claim areas in the Arctic. These countries are eager to make use of the natural resources. They do not want to rescue the North Pole! Apparently, to them global heating is not a big concern;
- Every year one particular country organizes the 'Bilderberg Conference'. This conference was founded by Prince Bernard of the Netherlands. The richest, most powerful and most influential people in business and politics have an 'informal' meeting. There are no formal records of these meetings. Still, it is a well known fact that important agreements are made to protect and strengthen the interests of its participants;
- The former head of the Security Service of the Russian oil group Yukos has been sentenced to 24 years of prison for organizing assassinations in order to serve the business interests of Yukos. Alexej Pitsjoegin was tried trail along with five others people who were sentenced from 7.5 to 19 years in prison[5].

The notion of dangerous people often brings stereotypes to the mind including criminals, murderers and so on. However, we know that there is more to it. I would like to conclude this paragraph by stating that dangerous people are the ones who put the survival of the Earth at risk by maintaining a system which is based upon growth rather than the balance for all life on Earth. Their only goal is to gain more and more in a short term. Think about it!

Distribution of wealth

I would like to stress that I do not have negative feelings against rich people. I wish everyone a prosperous life. People should be allowed to enjoy the nice things in life. Especially, when they worked hard for it or thrived by the development of a brilliant idea. If I had a lot of money myself I would probably enjoy it too. Still, I do not wish to complain and will not do so. It would just be more fun if there were more people with a happy and wealthy life. It is like the state lottery. The jackpot continues to rise every month. This is done for commercial reasons to show off the biggest prize of all lotteries. For example, a person wins 20 million Euros. For the same amount of money 20 people

5 From Nu.nl August 17, 2006.

could have gained 1 million Euros each or 40 people 500,000 Euros. Thus, the same amount of money might have made 160 people (= 4 people x 40 families) happy instead of only 4 people (1 family).

This mechanism goes for the entire world. A relatively small proportion of the world population lives an extremely good life. Let's assume that this group consists of approximately 1.5 to 2 billion people (North America, Europe, a part of China and small parts of India and Russia plus another couple of hundred millions in the rest of the world).

That is a huge amount of people one might state but there still remain at least 4 billion people whose main concern is survival. Often, the happy few gathered their wealth at the expense of the poor. Their wages are low and the rich bolster this situation in order to export cheap products from second and third world countries to the first world.

2.5 Unfaithfulness

Marriage

Do we still understand the meaning of faithfulness? If one examines the number of broken marriages and relationships and the number of partners cheating on each other it seems that we have lost track. First a man marries a beautiful woman but after some time she becomes boring and that is when he decides to have a new, more attractive wife. This does not sound very pleasing but I would dare to contend that people nowadays maintain a relationship as long as it seems to fit their age and state of development. Every phase in life requires the right partner, one who is able to keep up at the same level even if this means being involved with a new partner. This notion has become a common idea. A divorce is painful and it has a huge impact on the development of children. Many people do not realize beforehand just how sad and sorrowful a separation will be. They experience the tragedy of breaking up their marriage, only so much later (they do not listen to the elderly, remember?). On the spur of the moment one believes his own development to be more important. Thus, he or she chooses to go for his or her own sake, regardless of the consequences. Yet, this leads to less and less profound relationships, ensuring that if the relationship was to end we would not be hurt. As a matter of fact we pass this way of thinking concerning marriage and relationships on to our children. As a result they will maintain only superficial relationships as well.

The family
A family is a micro society. How we raise our children has an effect on the macro society in which we meet various kinds of people. This means that a less cohesive family containing superficial relations will reflect its image upon society in general. Is this really what we want? Does that make society a better place?

Businesses
The matter of being loyal to the man in charge or to the company itself has gone into oblivion as well (or is this just the result of the above mentioned issue?). You work for money and for your own welfare not because you are proud of the company you work for. The next day the company might not need you and it may fire you leaving you empty handed. Therefore, everyone ensures they are a valuable specialist or a widely deployable generalist (manager?) in order to switch from company to company to earn even more money. Thus, society within a firm adopts family forms.

2.6 Religions

As people we believe in all kinds of things. Just think of anything you can imagine. There are however a number of key trends within our society. The main beliefs are the regular religions, namely Christianity (mainly among Catholics and Protestants), Islam and Hinduism. Buddhism is also an important religion, but rather a belief in the meaning of a way of life. If you look at the essence of these religions, you will notice the same basic principles. Only Hinduism is somewhat different, because it is the only one with multiple gods. In Buddhism there are also gods, but they play a completely different role. It is about the teachings of Buddha.

Focusing on differences
If we were limited to the main features of the religions, then there would be no problem. But what we do is emphasize the differences. We all want to be right. Instead of combining forces and using that for a good cause, we continue fighting each other. We are lacking respect for each other as people.

An example: the Jews, Christians and Muslims actually all believe in the same God. Hence Jerusalem is the centre of all these religions. All three know the

Torah. For the Jews the Torah contains the law of God, Christians know it as the Old Testament from the Bible and the Muslims have also picked it up, but they say that the old Jewish priests have changed too many of the original words. As a result, Jews and Muslims are having much the same festivals and memorial days. When Jesus Christ was born, the Christians saw him as the son of God. The Jews had another point of view and they still do not recognize Jesus as the son of God. For many, he never even existed. Muslims see him as a prophet and not more than that, just like Mohammed. Since then Muslims and Jews have paid more attention to the strict observance of all kinds of rules and laws, because that is their way of heaven to come. Christians on the other hand believe that Jesus died on the cross for their sins, which freed them of strictly complying with all sorts of rules and laws. All your sins are forgiven afterwards, if you just convert to Jesus Christ. Who is right?

Another interesting aspect of the Torah is the subject of consulting ghosts. This is strictly prohibited. I am not suggesting that we should consult ghosts, but what can we learn from this fact? Apparently most religions also believe in ghosts, dead people whose soul is still hanging around on Earth. It will add to the image of the universe which we are looking for.

Power versus love

There is also another problem, that most religions are abused by a small group of people to make themselves important and able to exert power over others. This is mainly done by making religions more complicated than they are and by claiming that they are the only people who know the true facts about the religion. This means that they are the people to whom everyone should listen.

When I look at this picture about religions, I strongly wonder whether this is the real intention? Is love for each other, including yourself, not the essence?

If there really is a 'God', to whom love and goodness are of most importance, who wants everyone to come to him, I cannot imagine that he would make his teachings so complicated that it needs study or discussion. An intelligent being, the highest in the universe, which does not care who you are, has no need to excel by complex writings. Such a supreme being would rather see that his creations are happy and love each other. This is for everyone to understand.

2.7 Dealing with facts

Scientists should continue to help mankind with change and try to protect against mishaps, whereby people, animals or the Earth could be damaged. Regarding the medical science we can say that a lot of positive things have happened over the last hundred years. All kinds of vaccines against serious diseases have been discovered. We can really prevent or cure many diseases. We are also able to help many people by surgical interventions and with prostheses. And the developments still go further. We are able to give people more kinds of help through our health care system.

Nuclear power

It is a shame that it needs a lot of sick and/or dead people and much scientific evidence, before we accept that something is harmful to humans. We do not start with scientific research. Business comes first. Then we start looking for the consequences. And although most of the time we already know if something is harmless or not, we wait until it is an established scientific fact.

If you work at a nuclear power plant, you will receive a lot of radioactive radiation. As long as this stays within certain limits, it does not appear to be harmful. Radioactive radiation is not healthy. One body can endure a lot of it and another cannot. This depends on so many factors, that it is almost impossible to predict the outcome. By working in a nuclear power plant you run an increased risk of getting cancer later in life. Is this worth the risk?

We also often ignore the research results in the importance of business. In England you can find the nuclear power plant of Sellafield. Many children in the neighborhood get leukemia. A huge amount of radioactive material is dumped into the sea or finds its way there during malfunctions and accidents. The British government says that the studies prove nothing and that the leakages do not hurt anyone or anything. Are we doing a good job? Do you think that many countries want to expand their nuclear program? How hypocritical can we be? Tomorrow there may be a nuclear power plant in your backyard.

Have you ever looked at television programs concerning children with severe forms of cancer? Children have a wonderful innocence, not knowing about all the evil in the world, all the pain that is suffered. Children are so pure. There is nothing more beautiful than the smile of a small child. A child wants to play, does not want to worry, because it does not know what that means.

But a child with cancer lives in another world. His or her youth is immediately over. Such a child lives in a world of pain and concerns and sees the tears in the eyes of his parents. That is really terrible. It is already difficult for me to write this part without getting tears in my eyes. I understand that we need energy to live, but if that same energy destroys the lives of small children, then ask yourself whether we are on the right track. And I'm not even talking about the nuclear weapons that we make with the help of nuclear power plants. There are still people in Hiroshima and Nagasaki having pain daily and people dying, as a result of a few A-bombs from over 60 years ago.

Mobile phone and smoking

The use of the mobile phone is a good example regarding the lack of scientific proof. Various studies have already shown that the use of, or merely an active, mobile phone in your pocket has physical effects. Until now, we cannot prove that these effects are harmful. But who wants to be subject of a commercial experiment?

It is already known that not everybody can endure the same dose of anything. Substances or radiation not harmful in small quantities to one person can be harmful to another. Meanwhile it has already been proven that smoking can cause lung cancer. Yet there are people who smoke throughout their lives, eighty years, and do not get lung cancer. Smokers take people like this as an example. It is not wise to raise exceptions to a standard.

Military purposes

A lot of scientific knowledge is being misused for military purposes. A good example is the development of the atomic bomb. Knowledge of atom splitting, intended for generating energy, was in the Second World War used to develop the atomic bomb. Hundreds of thousands of people were killed in Japan, while the nuclear scientific knowledge was developed for more positive purposes.

Rocket science, initially intended for space exploration, is very often being used for military purposes. Everywhere around us swarm military intelligence satellites through space, which gather information about anything that possibly connects to military activity. The missile technology is more often used for military purposes than for exploration of the universe.

Abuse and self-interest

But the worst thing of all is perhaps, concerning honesty and truth, that people misuse facts, research or by referring to it, to make a story look logical and acceptable, while it is not, just to reach a goal. A good example is the fact that political parties in the Netherlands consider that the aging of a large part of the population will make the old age pension priceless and weaken our economy. Years ago a previous government had already seen it coming and there was a lot of money saved to keep the old age pension affordable. Since then no one heard anything about it. If they had kept on saving more money, we would not have had this discussion now. Another fact that we do not mention is that the aging of the population will be ended in 30 years or so. It is only a temporary problem. You may also choose to let the state debt rise temporarily and resolve it later, just like a normal household does that also sometimes has temporary setbacks. If you do this, then there is no need to make the working class of today pay for something that was badly arranged in the past. These people are now expected to work longer and also to pay more taxes for their pension. They will pay more for their pension than the retired before them, but they will get less in return. Is this unfair or not? Political parties want to score when they are part of the government by showing a clean budget during their period. Big business companies are lining up to give the politicians a well paid job afterwards: own interests over peoples interests?

The environment

Maybe it is not the right solution to temporarily allow the state debt to rise. But most important of all is that we leave out more important facts against further growth of the economy: the scientific results on the environment: things are going the wrong way. The pollution of the environment is becoming worse. The climate change is going faster and faster. We are always extending the boundaries of what is acceptable. We say that there is still room enough for the economy to grow. Are we deluding ourselves? Look at many metropolises such as Tokyo and Beijing, where people walk the streets and cover their nose and mouth, because they fear the polluted air. This cannot be healthy? The Netherlands has a limited size, limited infrastructure and is one of the most densely populated countries on Earth. At the time the mobility stops, the economy also stops. More and more cars are increasingly making use of the cancer procreating diesel, making many countries inhuman to live in. Cancer is already the number one cause of death in a country like the Netherlands in 2008.

Logical thinking

Key fact is that we are pulling a heavy burden on the environment with our economy, which is changing the surroundings we live in. Slowly we are destroying the ecological system. Animal and plant species are endangered, we fish the seas empty, the trees are felled everywhere. All kinds of statistics, such as average temperature, CO_2-level, fine dust measurements, etc. show that our economic growth cannot go through in the same way. We are clinging to the current system and are taking inadequate measures to slow down the growth of the world population. So now we are heading off to the downfall of our own species. Why don't we intervene? Because the people with the power only look at the short term: it will last my time, if my budget is all right now or if my capital is growing everything is all right, because I want to go down in history as someone who had his business under control and doing better than the rest. The multinationals want to see the earnings rise, because that is what their investors and shareholders are asking for so that they can fill their pockets and enjoy what is left to enjoy. Why statistics? There are scientists who say the opposite! Predicting the future is very difficult and there are many factors that affect the ecological system. But is this is not some kind of stupidity, against any form of logical thinking? On television shows years ago it was funny to see how many people were in an ugly duck (Citroën 2CV). However, at a certain moment the fun stopped, because there is a limit. The same applies to the Earth. Is it wise to search for the limits of the Earth and beyond? Is it necessary to risk everything?

Economic situation

At the beginning of the 21st century the new government in the Netherlands said that the economy was not doing well. Furthermore they said the economic position of the Netherlands in Europe was not good. They blamed the previous governments (it is always easy to blame your predecessor, because they will not get the chance to prove otherwise). At that time I stumbled on an article written by an economist, who said that there was a global economic downturn which caused most of the problems. Furthermore on the basis of international figures he showed that the Netherlands were not doing bad at all. But these figures were not shown by the Dutch government. Indeed, according to official figures, the Netherlands had in fact a very good position in the world on various social and economic areas. Anyway, if you show such figures, you will not get the votes you need to be able to make changes. I do the same, but I have a really easy job, because it is impossible to close your eyes for all the misery in the world.

Then the economy in the world recovered slightly and the new government told us that we had to thank them. But of course that was not true either.

The big companies are increasingly filling their pockets, mainly through tax cuts and export abroad, but not because of rising sales in the Netherlands. The Dutch themselves are increasingly lowering their savings, debts are going up and many companies have large supplies. The situation has not changed that much.

And so we are always trying to show a picture supported by facts that we have gathered ourselves. It is not based on the most relevant facts and therefore not the real picture. Thus we are fooling the people.

Employment

At the end of 2007 the Dutch Central Planning Bureau proclaimed that the economical rise of India has positive effects on the Netherlands especially. They said that India is a country of products for which mainly low-skilled labor is needed. What a mistake! Don't they read the newspaper at the CPB ? India is now one of the greatest performers in the field of ICT in the world. Millions of jobs from the United States and Europe have now disappeared as a result of outsourcing of ICT services to India. No one protested! And the expectation is that more of the services will go the same way. In the United States more and more middle class people are having difficulty finding a job. Again, this is another example where the government authorities do not show us the complete set of facts and give us a false picture of the world.

Intuition

A very popular word at the moment, used in all kinds of contexts: taking decisions from your intuition, because if you deal with your own feelings in the right way, you can do this. But aren't we forgetting something? Many people are not balanced and are driven by fears and desires. If you are also driven by money and power and not by love for yourself and your fellow men, then making decisions from your intuition is dangerous, especially if the decision also affects other people. This leads to taking decisions that have not or hardly been substantiated with facts. Purely logical, this cannot be justified? Or is this another attempt to make people stop thinking for themselves?

2.8 Responsibilities

Self responsibility

Taking responsibility for ones actions is another interesting problem in our society. A liberal term widely used, especially in business and politics. The argument, which is used to stress its importance, is that they want to prevent people from becoming lazy and dependent. Independent people who take responsibility are more productive, more creative, etc. A nice thought, but on the other hand they are expecting us to sign all kinds of insurance policies, because otherwise we are considered irresponsible! Now anyone can understand that this has nothing to do with self responsibility. There is something else behind it. When you close an insurance policy, someone else takes responsibility when you do something wrong. How is this self responsibility? We are always trying to let someone else pay for our damage!

The reason that politics and business are talking about personal responsibility is the fact that they do not want to take responsibility for others and want to save on costs! The less responsibility they have, the fewer premiums they need to contribute to cover the risks. A good example is the regular discussion whether someone should get a salary, if he was injured at the weekend during a game of football and cannot go to work on Monday. The employers prefer that everyone insures themselves, so that they do not need to pay out if something occurs. On the other hand however, the employers are forgetting that they also want their employees to do some kind of sports, because such workers are generally healthier and more relaxed, making their performance better. And since everything you do causes risks....Someone who is just sitting on the couch and does not do any sport also has an increased risk of certain diseases. Do we have to be insured for that too?

Example Function

It is an increasingly popular thought in the Netherlands that everyone should be able to show who he or she is. This is not about freedom of speech. This is about people who have a lot of money and who want to be able to show off. Apparently the people about whom we read in the papers and look at on TV, the VIP's such as directors, managers, shareholders, politicians and famous people do not realize their responsibility for behavior in relation to the 'ordinary' people. Because if you, as a well known famous person, start throwing money away, buying energy-guzzling houses and cars, then many other people think that this is the norm.

If you are a politician calling for people to save on fuel, you should not go driving a petrol wasting limousine or go to formula one races even if it is in your own free time. Or increasing your own salary by ten or twenty percent, while asking the population for wage moderation. Then you should not have become a politician. That also applies to the 'greed culture'. If directors start filling each other's pockets with huge amounts, it encourages other people to do the same. Why should ordinary people be satisfied with less? You can try to justify this, but it is in fact irresponsible behavior. If people look up to you, they will start copying your behavior, whatever behavior that is. If you want people to show positive behavior, you will have to show that behavior yourself first, or it will not work.

Kindergarten

Another example of the shift of responsibility is how we treat our children. We let them come home at noon from primary school, while nobody is at home. We give them the responsibility of a house filled with all kinds of stuff, which together are worth hundreds of thousands of euros. An accident at home can easily happen, because most accidents in our society happen at home. The child can die in a fire which started in the kitchen, while trying to light the stove. In addition, the child can have an accident while traveling to or from school or be picked up by some stranger. As a parent you are still responsible under all circumstances, because the child is a minor. In spite of the fact that the child could also have stayed at school!

Another example is children who must look after other children. What if something goes wrong? You cannot blame the child who is looking after the other child, because he/she is not an adult. You cannot expect appropriate action from a child in the instance of serious problems, since he/she is not educated or trained for such situations (are you?). And what do we do all this for: to save costs! Are we just thoughtless and easy or do we know better?

Frankly, to me it seems very noble if you are willing to take a certain degree of responsibility for another person. People then feel secure and safe. A nice feeling, isn't that what we would like to offer people? Or do we want a society with every man for himself, where you are immediately left out, when you have failed to take your own responsibilities? Do we want to help these people instead of just punishing them?

2.9 Individualism

We have become indifferent and turned in ourselves nowadays! The number of various communicational resources and opportunities are enormous, as well as their uses, but that is not because we find it important to communicate with other people, but because we want to feel important. We have all these resources, and hold someone on the line, who listens to what you say. That gives a sense of power, of self-affirmation. Actually, it is largely irrelevant what the call is about or what the other person says.

Private interest

The individualism or should I say the self-centered reign's supreme in combination with materialism. This all sounds terrible, but in effect that is what it is, for instance people are being killed for a portable radio with earplugs. We live close together, but we hardly talk to each other or greet each other when we meet in the streets. And if someone is beaten up or mugged we do not intervene, because we might get into trouble. Or if someone almost drowns, then we do not want to get wet, because our mobile phone could get damaged. Nowadays we think about our own interests first. When we are looking for new friends or a partner, we first look at what this person can do for us before we let him or her into our world.

In modern relationships we are primarily interested in what we gain by it. As long as the relationship helps us to develop, we hold it in position. What happened to the period when we said there is nothing more beautiful than to give?

The government is not making things easier by stating that everyone should have their own income so as to ensure their independence. The fiscal measures have been updated to support this way of thinking. The fact that this individualism goes at the expense of family life and the upbringing of children, is of secondary importance.

The capitalist system

The capitalist system leads to an ego-culture. The concept of 'together', when we talk about our society, has a financial meaning only: you pay taxes to maintain the society as a kind of company.

The modern capitalist society is focused on the ego-experience, which emphasizes the egocentric feelings in humans. Proponents of this idea will say that it is not the fault of the capitalist system, but a logical consequence of the development of people within society as a whole and that it is fairer to the nature of people. You can however ask yourself whether you are not just emphasizing one of the weaker sides of mankind? Because if it is every man for himself, what remains of the principle of society? If everyone just wants something out of the system and puts nothing into it, as the rich are already doing (a lot of them are using all sorts of tax advisers to ensure that they pay little or no tax), what is the point in a society?

On the other hand it has been proven that exposing people to negative and bad experiences causes many people to behave in the same way. The capitalist system is a system of the law of the jungle. The competition principle is the starting point. This means that there is always a winner and a loser. There is no tie, only temporary.

To behave modest, honest and nicely in a capitalist system leads to the situation where you are in the back row, increasingly being suppressed by people with a big mouth and people who know how to handle creatively within the rules. You must be cunning and faster than others to make it in this society.

Is this what we want? A system where people jostle each other? Is this fair? Is this the appreciation we want to give to each unique human being?

The communist system

Communism is often mentioned as an opposite example. But that has not worked either (although the Russians and the Chinese will not agree: more than one quarter of the world has a communist system!). That is only partly true. A good plan that is executed by 'bad' people will always have a bad result. I have put 'bad' between quotation marks, because I mean that it is generally very hard for people to handle power and wealth in a correct and honest manner. And that is what also happened to the countries with a communist system. The former and current leaders were not coping with the power entrusted to them. On the other hand, we should not forget that they also had to deal with a lot of opposition to their system, mainly supported by capitalist countries. This made the leaders very suspicious, causing them to take very unfriendly measures against their fellow man sometimes. China is rapidly becoming the most powerful country in the world: what do you mean by communism is over?

Communism has good principles, but countries with a communist system as a starting point, often seem more like a dictatorship with a capitalist core under it. Money and power also play an important role. People are stowed away in tower apartments and work for the glory of the communist regime. There is little justice done to the individual human needs, such as differentiation and self-determination. Lately things are a little bit better in most of these countries, because they are slowly joining capitalism. But certainly in the past most people under a communist regime did not have the resources and opportunities to make choices for themselves. We should not forget that many people within our society do not have opportunities, simply because they are lacking training and/or money and therefore cannot get a good job. When you do not have sufficient intelligence available, then your options are limited. But we are still talking about real people, who also have dreams, and wish to develop. In addition, our society has so many laws and rules that you must wonder how much autonomy do we really have?

Another accusation against communism was that the people were poor and the rulers were rich. Isn't that the same in our society? However, in western countries people say that you only have yourself to blame if you are poor.

The rules in our society are created by the parliament and the government that we have chosen ourselves. Most measures are expressed with money and if you have enough money, then you won't have to bother. And if the rich are affected, then they directly contact the government itself, because they are accustomed to that level of communication.

Are we were really against communism or do we just pretend to be? Or are we just hypocritical? We are doing business with the Chinese and the Russians and at the same time we accuse them of violating human rights and we abhor their social system. Do you really think that they will take us seriously?

The role of religion in society

In addition, there seems no other objective of our society than the creation of wealth, but for what? What do we want to achieve with wealth in the long term? The Muslims have made an attempt by linking religion and politics and fully integrating religion into daily life. They have done this by making life subordinate to the Islamic religion instead of making religion to be a tool in life. They only know one way and try to make people live this way through all kinds of laws, by which many people, especially women, feel restricted in their freedom and they sometimes even experience humiliation. I can hardly imagine that a 'God', who says he is love, wants people to be so limited in their possibilities. But perhaps I am wrong.

Christians believe that all their sins will be forgiven if they ask their God for it, because Jesus Christ died on the cross. So if your life hasn't been totally 'straight', it is not so bad. Everything will still be alright. I think this is just a half-truth. Maybe your sins are forgiven, but all the problems you had in this life have not all of a sudden been resolved, because you are dead. You will still have to learn from the mistakes you have made, so it will undoubtedly have consequences. A lot of information about this is available from the teachings of the Buddhists, hindoeïsten, paranormal gifted and near-death-experiences. And I may be wrong here also, but try to understand what I want to say: collect all knowledge, try learning from it. It will bring you further, it will bring humanity further. The individual should not be subordinate to religion. Religion is meant to help the individual.

And so it remains muddling through fragile social systems.

2.10 Unnecessary complexity

Less rules and laws

Another important issue concerns the lack of simple solutions to problems in our society. We make everything unnecessarily complicated, so people no longer understand how things are working. The society is becoming incomprehensible to many people. That is why you can hear politicians and intelligence agencies increasingly saying: 'Do not let us involve the people, they do not understand' or 'otherwise they will panic'. By making things complicated, there are also all kinds of agencies and officials that need to work with these complex issues. Only big companies can afford to pay for specialized consultants. This gives them a feeling of power and you can earn a lot of money by doing this: take a look at lawyers, notaries, accountants and tax consultants. Is it really necessary for the (tax) legislation to be so complicated? Why do we give people subsidy first and then let them pay taxes? Or simpler: why do we have a gross and net salary? Why do we have to pay taxes on all sorts of things and at different rates? Actually, with an extensive administrative procedure you may wonder why all this should be so complicated. The reason is actually very simple: 'because we have chosen this ourselves!'. We cannot shove the responsibility for our way of life onto others. We ourselves agree to everything being so complicated. If we want everything arranged differently and simpler, then we will have to start working on that. Choose politicians who really want to do something about it and who do not make things up to win votes.

Less traffic jams

The same applies to problems such as traffic jams and smog formation by traffic. We make all kinds of complicated solutions like large cargo railroads from Rotterdam to Germany, parts of a highway only open between certain hours, electronic toll roads combined with GPS, for which very large investments have to be made. And when they are finished, very often they do not even work properly! Just a lot of wasted money and energy. Is the real problem not inherent to the system that we use? We need more and more people to go to work to keep the economy running, thereby increasing mobility, which causes problems. Firstly we will have to ensure that there is no need for more and more people to go to work, so the mobility is no longer growing. Then you can come up with definitive solutions to traffic problems. Always search for the cause first and then devise a solution. With only symptom control, you will never fix the problem permanently.

Disease prevention

Maybe we should state things firmer still: if something is very complex, you may assume that it is not going to work, or at least at some point it will go wrong. That means that if something is very complicated, it is better to go back to the drawing board and see if there is another way to a simpler solution, which is indeed going to work. A good example is the study of diseases. This is a very money consuming business and we are making a lot of progress in many fields. We can go on giving more and more money and build the most complex theories about certain diseases, but it is probably much cheaper and perhaps much simpler to prevent many diseases in the first place (this usually means not doing something, which is generally easier).

The road of love

And the best example of this is perhaps the universal way: the path of love. This is what it is all about. What is complex about love? Everything you read about religions, near-death experiences, etcetera, you name it. They all come back to the same point: love yourself and others, it is that simple. There is no excuse for not being able to understand that. This message is for everyone and anyone to understand. For me, this is a clear sign that the solution is simplicity.

2.11 Economy and money

Foreign investors

The multinationals have been given lots of opportunities to do business in an unequal competition with the low-wage countries. Many Dutch companies have now been sold to foreign investors, who are now playing ball. As long as the taxes are low enough, they will keep doing business in the Netherlands, but as soon as there are cheaper alternatives, they will leave without a tear, because they have no cultural or emotional relationship with this country. That means, however, that the wages in this country will remain under pressure and that this country in the long term will be unaffordable to live for the less wealthy. One of the reasons is that the government must also get its money from somewhere. If it is not from the profits of the companies, then they must get if from the people who live and work here.

Thus the middle class is under considerable financial pressure from two sides. I am not talking about speculators, investors and investment companies, who will push the prices of food and fuel up to earn a lot of money in these markets too. This will lead to a situation in the long term where there will be little left of a middle class. So if you are not a business man, director, manager or advisor of these people, then you will have difficult times in the coming years due to the need for low wages.

Small businesses are also having a difficult time. If they are not pushed out due to competition from the big boys, then they have to deal with the increasing regulation from the government.

Other major disadvantage of companies in foreign hands is that these companies have no need to invest in our society. Next year they might be doing business in another country. They are black holes in our society. And just like the black holes in the universe they pull everything inside and give little or nothing back. In the long term this can have disastrous consequences for our society, because what if there is nothing left for them to get out of our country? And the more black holes there are, the more likely that this situation will occur.

And what about the control over our own country? Not all the important measures and decisions have an economic reason. Sometimes there is a cultural or social motivation behind it. But foreign investors are not interested in these kinds of things. Therefore they can use (usually invisible) pressure, causing such measures and decisions to be taken less and less. The Netherlands is increasingly becoming a subsidiary of a foreign company, whose inhabitants can only be seen as (potential) workers and not as participants in a society with its social and cultural aspects at which it is necessary to pay attention to.

European Union

The introduction of the euro in 2002 and the millennium problem caused a lot of work in the Netherlands. The economy was doing very well. But the terrorist attacks and the huge price increases as a result of the introduction of the euro caused a huge economic downturn and almost a recession. The purple coalition, a combination of left and right parties, which the Dutch people welcomed at first, was voted out after two government periods and discredited. There came a right-wing government, under the guise that once again, we should respect norms and values, work hard and no more lying on your back with money from unemployment benefits in your pocket. But in reality, this is the result of the ongoing globalization, in which we increasingly have to compete with low-wage countries. We therefore need an even higher production rate at lower costs. In fact, this is a fight we will lose anyway, but because we now have a system of open world economy, we seem to have no choice. However, in the long term it means we will have to go back to the beginning of the industrial revolution, when factory workers were still really poor, just the same as in the low-wage countries of today. In this respect we can learn a lot from the United States, where this process is already much further. Many former factory workers and people from the lower half of the middle class find it increasingly difficult to find a job. This means that in the long term, the middle class will disappear and only a rich and poor class will remain. The situation about foreign investors will contribute to this.

We are still (2007/2008) scramming for workers, because the economy is at a high. Many people who are now unemployed or disabled are in fact written off and will never get a vacant quality job. If there is another recession in the coming years, which is unavoidable in the system that we know, the number of unemployed will rise quickly. What will we invent this time to keep the unemployment and disability rates low? In a system in which production should be increasing all the time, you will always need 'stronger' people. Fewer and fewer people will be able to comply (over a longer period of time) to the ever increasing requirements and will finally drop out.

Many people will say this is far too negative. Look at the number of people that have got a better life than ever before! And indeed, I cannot deny that the average level of prosperity has increased, but there are also more poor people than ever before. What is the price of this increase of wealth? Let us confine ourselves to the economical aspects. Previously only the man in the family was working 6 days of about 8 hours, which is 48 hours a week. The woman was looking after the household and taking care of the children. Nowadays the adults in a family are both working, so that equates to approximately 60 to 80 hours a week. In addition, more and more social services are broken down, causing people to handle setbacks for themselves. The people who are smart and healthy are economically doing very well. It is a kind of rat race, where the weak and elderly people are increasingly dropping out. And remember, even if you win, you are still a rat!

Taxes

Every citizen gets annoyed over the taxes. A system that has come up fairly evenly as far as the development of money is concerned. The first tax laws of the Netherlands date back to around 200 years ago. Since that time more and more tax laws were made. There have been attempts to reform and simplify the tax laws on a regular basis, but in essence it has never really succeeded. One of the reasons is undoubtedly that if the ordinary citizen would understand the complex tax code, they would probably see that a lot of taxes are not right or even obsolete. Elimination of these taxes would mean a huge problem for the government budget.

On the other hand, there is a lot of time and money invested in the tax-system by the government, tax-advisors, information and computer technicians, and of course everyone who must fill out tax forms. Time and money that could actually be spent in a more useful manner.

In addition, companies are increasingly paying less and less taxes and citizens and paying more and more. The number of citizens in financial trouble will be growing. One good example is the Corporate Income Tax, which was reduced in recent years from 40 to 20-25% depending on the level of profits. The dividend tax will be reduced from 25 to 15% in 2008. These are just a few of the changes made by the government that are favorable for companies and shareholders.

Safety versus economics

Everything is in favor of the economy, even at the expense of people's safety, how else can you explain that cars such as SUVs[6], which are dangerous to other road users, are allowed? In addition cars that have more than the average fuel consumption, while we are trying to diminish the pollution pressure on the environment? Why do we allow cars with white flashers, making it difficult to see if someone is indicating to go left or right? Why do we allow cars of which we cannot change the lights ourselves so that people drive around with a broken headlight longer than is actually allowed? Why are there xenon lamps in cars, which blind people on the road, and thus endanger traffic safety? Only for the economical importance! Is this justifiable? Who dares to say that we are still doing a good job? We take no logical decisions. We are unable to differentiate between major and minor issues. We do not choose people as a main issue, but we choose for the system instead!

Rich versus poor

The rich in society continue to do just what they always did, because they have the money to pay everything off. Even when it comes to environmental issues. Extreme petrol-consuming cars get an extra tax-charge, and the excuse: "I paid for it, didn't I?" can be used. Moreover, the extra tax-charge makes these cars even more interesting, because only the really wealthy can afford them. And do you really think the rich are likely to fly less? Do you think they are ever going to use as little energy in their homes as you? That they will use as little fuel as you? Why do they have the right to use so much? Because of evolution, because of the past? Because they have shown that they are more important than other people? Do not forget that wealth is for a large part the result of luck or origin. Even intellect is for the biggest part the result of genetics and environment. Most people work hard, but do not get rich.

6 SUV = Sports Utility Vehicle, is a semi-off-road car for normal traffic. These cars stand somewhat higher above the ground, compared to normal passenger cars. The bumpers are also somewhat higher, causing these cars to slide over normal passenger cars and to completely run over pedestrians in an accident.

We humans have been given brains and the choice to do what is fair and reasonable and to be able to think of how to change the world to our advantage, if we want. The rich and powerful are equivalent to every other man and they should behave accordingly. The evolution is often used as an excuse of why we behave the way we do. This seems a correct answer, but haven't we also developed ourselves into rational thinking beings, allowing us to make choices? Therefore the problem not only lies in the system, but also in human nature itself.

In 2007 The International Monetary Fund (IMF) concludes in a study that the inequality in income has increased over the past 20 years in the entire world almost. The capitalist system is therefore only to the advantage of a minority of the world population. A couple of beautiful examples are related to how bank companies work. If someone in the United States wants to borrow money at a bank and that person is important for that bank, because he has a lot of money or much influence in a given area, then that person can borrow money against an interest rate of 0 %. On the contrary, when it is someone, who really needs to borrow the money, because he has no collateral and no goods, then that person often has to pay double the normal rate. In short, that person has to pay more money to borrow the same amount. Another example is the stock market. If you have a lot of money, you will get a personal advisor in a separate department of the bank, who is going to manage your portfolio. Because there is so much attention and time spent, the return is generally much higher. However, if you have less money, you will have to do with equity funds or even buy shares by yourselves. Because you only have limited time and knowledge and the equity funds in general consist of an equal mix of shares, the return that you make is usually much lower. And if you put a lot of money into a savings account in the bank, you get more interest than someone with far less money on his savings account. Whatever you do, if you have little or no money, your options to make progress are absent or very limited.

Economic system

If there are problems in society, then they can be generally related to the financial and economic system that we use. The economic system with its growth principle justifies on the one hand the pushing of people to work harder and harder and on the other hand the growing world population, the cause of most problems. But if we do not push people to work harder, nothing happens, many will say. But who are you to determine whether someone is working hard enough? Have you ever really taken a good look at the situation? Probably not, because you are also in a system where there is not sufficient time to look carefully at things. Perhaps there is a sound reason why that person is performing the way he or she does? But there is usually no time to look for those things. And on the other hand, what need is there to make people work so hard? Where are we going?

If the economy grows, jobs are created resulting in a growing need for workers. The population must also grow in numbers. The economy usually cools off, leaving an increase of the unemployed (because the number of people does not increase). As a result the economy is pushed forward to create jobs. In this way we create a system which results in a vicious circle. We are continuously trying to solve one problem with another problem.

Financial system

The use of money was actually intended as a tool (for trading, payment tool, for accounting and as a saving tool). You do some work or you make a product and if you sell it on the market, you get money for it, allowing you to buy something else of equal value. In this way we tried to create a certain balance. Gradually we saw that by owning money you also obtain some kind of power. The link between power and money has provided a different view on money: a power tool.

Money can now be created by itself, without having to do much with it, for example by making it available to a bank or by buying shares. The purpose of the use of money in this way is no longer as a tool but as a purpose in itself. In society money has become something for which it was not intended. People gain power over people and resources by having a lot of money, which is not always based on hard work. Thus there is speculation on futures and options on food commodities and oil in 2008. Despite the fact that these products are available in sufficient measures, the prices go up. The market of supply and demand no longer works as intended. A nice, fair principle is finally ruined.

On the other hand we also see that the use of money is not a closed system. Frequently there is money created or destroyed, without something in return. For example, if poor countries or people have too many debts sometimes part of it is relieved or a company may go bankrupt. Or if shares are suddenly worth more money, there is nothing to balance that.

The use of money is not in accordance with its original purpose. It is used for other purposes which lead to the use of money in both a positive and negative way. This should end.

Why don't we abolish the use of money and go back to the original starting point? We only need something for people to obtain the necessary products to live and nothing further.

2.12 Overpopulation

We are increasingly confronted with the consequences of overpopulation and economic expansion during the last years, which in my opinion go hand in hand. Economic prosperity attracts namely poor people, making those areas overcrowded. We have done little on this issue in the last hundred years. Indeed, we believe that more and more people are needed for our economy to grow even faster! We therefore only increase these problems.

Multicultural society

Part of the problem of overpopulation and economic expansion is that societies are becoming increasingly multicultural. People from poor countries are increasingly going to wealthy countries and into the cities. On the one hand we will meet more cultures and learn to respect them, while on the other hand it makes society increasingly complex.

The Netherlands has become a multicultural society, where people are living close together. More than 17 million Dutch people were counted in 2006. The largest concentration of people being around the big cities like Amsterdam, The Hague, Rotterdam and Utrecht. We call this the 'Randstad'. There are also a lot of smaller cities that are part of the 'Randstad', but the big cities form the major part. Large parts of the big cities are inhabited by immigrants, this is a clear sign of a multicultural society. The Netherlands is no longer owned by the native Dutchmen alone.

The first large groups of foreign workers, mostly Turks and Moroccans (but also Spaniards, Italians and Portuguese), came to the Netherlands in the early seventies. But at the beginning of this century the government concluded that the integration of these people has failed. There are various attempts to conclude this in a forced way. Nowadays the workers from the former Eastern

Bloc are coming to the Netherlands. We see that the same pattern is repeated. They come with their colleagues and stay in small homes and get paid below the official minimum wage. Lots of them work here illegally. After a while some of them call their families to come over to stay here permanently. But they have a different culture and language. It is not easy to change and a lot of immigrants do not want to change. They are only here for the money.

In addition, it is said that everyone is entitled to his own culture and that we must respect that. What do we want? Do we want to be one cultural society or do we want to be separate cultural societies in one country? We do not really make a choice, we are unable to penetrate to the real problem. Is this typical Dutch or do we see this in other countries too?

However, we must realize that the more differences we 'allow', the more difficult it is to work together, to cooperate, you must also be able to understand each other. Almost everyone knows the concept Babylonian confusion of speech. Are we going to manage that risk?

Even in Belgium we see that it is impossible to live together using both French and Dutch. In 2007 they seriously considered splitting the country into Dutch and French speaking areas.

Traffic

The highways are full, even on weekends, and we must continue to take measures to keep life at the same level. Deterioration is almost impossible to prevent. In large parts of the Netherlands you can still find calm and space enough, but we have failed to economically develop these areas, making it almost priceless to do this nowadays. We will have to do this with the current big cities in the center of the Netherlands, but this part of the country is almost fully in use. Of course we will still have to accept that there are limits. What these are we do not know at this moment.

There are plans made to take care of the traffic jams, like new highways and extending the number of lanes on the highways, but in the end all those motorists need to enter the cities and that is where the actual traffic problems begin. And by stimulating the economy, the traffic will also increase. In that respect we are not solving problems, but only increasing them.

Another solution is being worked out: the electronic toll highway. This is meant to make people more aware of how and when they use the car. Hopefully people will start to use more economical means of transportation, which will lead to less traffic jams. But I do not think so. Only the people without enough money are forced to seek alternative transportation. Dare you call this social? What about the alternatives, such as public transport? The connections are bad and with a 5% percent increase in passengers a year, it would be a disaster, because public transport really cannot handle such numbers. The distances in the Netherlands are generally small, but to transport people in a short time to the place of destination remains difficult, especially because everyone wants to go to different places.

By stimulating both adults in a family to work this means that people have to travel more frequently and further than previously.

The use of diesel fuel is a separate story in this framework. Diesel produces soot particles that can cause cancer. It is widely known. This has probably already caused millions of deaths. But we have never considered banning diesel as car fuel. There is more diesel used and produced than ever before, and production is only intensified. The economy is again more important than people's health.

Possible solutions

We should build additional roads and houses in the populated areas. We build too few houses, because this causes the prices of land and houses to increase, which is in the interest of a lot of people like real estate developers and brokers. The real estate tax is making more money for the local governments of the cities. Side effect is that it also causes the rent of houses to double[7] and this affects a lot of people who do not have much money and cannot afford to buy a house, however this is not the main concern of the local governments.

Instead of building additional highways, we introduce the electronic toll highway, which will only shift the transport problems, but not solve them. The people still have to get to work. In a country like the Netherlands the creation of new highways is no easy matter, because the country is so densely populated. The construction of new roads always affects many people, who are against it most of the time. Everyone wants new highways, but not through their backyard. There are certainly more innovative solutions being studied, such as alternative fuels (electricity, bio-ethanol, natural gas etcetera) and building highways on top of each other, creating double the capacity of

7 The rented houses also increase in value, which in turn means the rent is also higher. But because there are too few houses being built, the demand exceeds the supply, causing the rent to rise even faster.

traffic on the same piece of land. That is a characteristic of man, we are always looking for new opportunities to cope with the problems in our world. It is important that we do not feel defeated by all these problems facing us in the world of today, but keep thinking about solutions in a positive way.

Forced Growth

The growth of the population in the prosperous areas is actually a forced growth, as a result of the need for labor power by the economic system. The autonomous population remains constant or is even diminishing nowadays. In countries where there is no economic prosperity we see otherwise. The population is growing strongly, because many people still have children in the hope that a few will survive. We send huge sums of money to poor countries, but things do not really improve there. The financial sums that are required are far too high. On the basis of money we will never be able to help these people. However, we could if we really wanted to, if we were going to help these people because we want to build a society out of love for our fellow men.

But aren't we really obliged to help, because Africa and India have been exploited for over a hundred years as colonies of Western countries. Only after the Second World War, these countries were 'liberated'. India is now slowly beginning to recover and is building an impressive economy, but Africa is still down under among others as a result of the AIDS virus that takes millions of victims.

The environment is degrading. One of the causes is the burning of the tropical forests for agricultural purposes and also for wood to be used in houses in rich countries. Another part is caused by air pollution from cars, planes and factories. All these things will only increase the problems in poor countries in the immediate future. The prosperity will remain low in these countries, and as a result the birth rate will remain high! If you're poor, giving birth to children is the only happiness in your life. Otherwise, there is nothing. A healthy child means joy and love. And hopefully it will live longer than you do.

Conclusion

Regarding one of the main causes of all problems on Earth: population growth, we are still not on the right track. Although meanwhile, we could have learned that in a wealthy community people have less children, the starting point of a balanced population.

2.13 Health Care

The health sector is an independent business in the current society. This means that wages and prices are primarily determined by the pharmaceutical industry, equipment suppliers, employees and freelance personal in this sector, making health care for more and more people unaffordable. In the United States there are more than 45 million people without health care, because they cannot afford health care insurance. Up until now this was acceptable! This is, however, 15% of the total US-population, which is almost 3 times the total Dutch population! If someone in the United States gets cancer and is insufficient or not insured, the only thing they can do is take an overdose of pills and a bottle of whiskey and say goodbye. This is very sad. We are talking about the biggest capitalist country in the world. A country that wants to be an example for many other capitalist countries, such as the Netherlands.

What about all the people in developing countries, which barely have any sort of health care system? While the people in the rich countries are helped by doctors for their wrinkles, excessive fat, for larger breasts or beautiful labia. This is all too crazy to even talk about! Incidentally, there are also too few doctors in the western countries, because there are not enough study places for young people to become doctors and a lot of trained doctors work part time. This is very strange, because we need more doctors and there are enough people who want to become a doctor. Regulating the number of doctors also ensures that the high salaries in this sector can continue.

2.14 War and violence

The history of mankind, as described in Chapter 1, is a past of war and violence. You would think that we would have learned by now? I wish this was true! Unfortunately, we are just going on and on making wars. Lots of innocent people have lost their lives, got injured or lost everything that they had. And the soldiers, who come back, if they come back, are often plagued by trauma caused by the killing of people and the sight of all the misery of war. No one stays untouched in a war.

Look at the wars of the last century: Germany started the First and Second World War, but ultimately lost both times. Japan started a war against the United States, but eventually also lost a lot and is still suffering of the side effects of the atomic bombs. Before the war in Korea, there was a North and

South and after the war this stayed the same. The United States sent a huge number of troops to Vietnam in order to stop the Vietcong, but all the violence was ultimately of no use. During the war in Yugoslavia, NATO was deployed against Serbia to stop the fighting there. The result is that NATO is still there. The same goes for Iraq and Afghanistan. In short, violence brings no permanent solutions. People do not support foreign intervention. Violence is only accepted in the case of self-defense. But once you try to let strange people under threat of violence do something, they do not trust you and it will not work in the long term.

And actually, this applies to all forms of violence. If you are the aggressor in the eyes of your opponent, then you can expect violence in return, and this will usually escalate. By using violence you break the existing balance in power. Once such a fracture occurs, the other party feels suppressed and will then take action to restore the balance back to an acceptable level, usually the level prior to the attack by the aggressor. A good example is the terrorist attacks on New York. The United States struck back by attacking Afghanistan. The Taliban was initially the aggressor and was punished by the United States. The United States went further down the line than before the terrorist attacks, through invading Afghanistan. At that time the situation changed because now the United States were the aggressors. This caused an aggressive reaction. It is actually very simple to analyze situations and behavior and then try to find a logical solution. The question is actually what is it all about: 'Why had the terrorist attacks taken place? Why were these terrorists asking for attention? "We have to look for a peaceful solution, there is no other option.

Why do we not learn? Why do we not find permanent peaceful solutions? Or are we in the Netherlands, crazy to invest so little money in our defense system? If the Netherlands join more military conflicts, they will lack personnel and material to send there!

2.15 Crime

Healthy people

'Mentally healthy' people find themselves in crime mainly because these people want to have something that belongs to someone else, and are not able to obtain that legally.

Poverty would not exist without money. Money can make people do all sorts of things, including sexual contact with a strange person or to murder someone on request.

Money and especially the need and desire to possess money, is leading to many problems in our society. Power and money are linked without a doubt, along with crime and money. Criminals are present wherever money can be earned.

Money laundering

There is also a very large group of people who are not named criminals, but who earn their money in a somewhat unfair way. In the Netherlands we call this money 'black money', because there are no taxes paid. That money should then be laundered, which is illegal, once again performing another unfair, and dishonest activity.

Without principles

Most people do honest business, we think. But there is something else to worry about. What about the honest business we do where we feel the need abandon our principles, for instance because it is in the interest of the economy or because there is a lot of money at stake? A good example is the business we do with China (the same goes for the business we do with a lot of oil-producing countries). This country is in fact a dictatorship. There is only one party, the communist party. They are in power and determine who governs the country. The people of China have no choice. Is a dictator, a criminal or someone who just knows what is best for the people and the country?

Furthermore, this country has an imperialist attitude. It occupies neighboring country Tibet, is driving the original inhabitants out of the country and has a hostile attitude towards Taiwan, of which it also wants to regain control. In addition, the human rights as we have agreed internationally are systematically violated. The rivers in China are 60% percent heavily polluted, leaving a large part of the Chinese population without access to clean drinking water. As a result many will have cancer. It will be of little concern to the government. And yet we do business with these rulers, even the Dutch government is taking part. We do that because we have an economic interest. "How straight can you be?"

Who is the thief?

The Netherlands is a country with an extremely high number of burglaries and thefts, mostly bikes. In October 2006 a newspaper reported that the Dutch police will give you a yellow card when you park your bike in the wrong place and in the worst case, the bike will even be confiscated! In other words, you must not only watch out for criminals, but also for the police, because you can lose your bike from either of them! Isn't that very strange?

Young people and immigrants

The criminality among young people and immigrants in the Netherlands is another worrisome development. This takes more violent forms than before. Even aid workers and the elderly have become targets. Stricter penalties have been put into place, and were discussed on television programs in 2008. But these are no long term solutions.

2.16 Drugs

One of the largest problems in our society is drugs, and I am not talking about the medicinal drugs, but hard drugs like heroin and cocaine. It destroys the lives of many people and the financial side is the foundation for a large part of the existing crime. Organized crime earns millions from the production and trade of all kinds of drugs. With this money they can buy real estate and businesses for money laundering. And the people who use drugs are slowly destroying themselves. Do not think: 'I can handle this', because you are mistaken. Cocaine is addictive and even if you use this regularly in small quantities, it slowly destroys your body. We have been trying to tackle this problem for years, but I read an article recently, the Dutch police admitted that we are losing the battle against drugs. In 2005 it became clear that you can get drugs in at least half of the Dutch secondary schools. What does this mean for the future of our children? They will increasingly be faced with drugs, while they do not ask for it themselves.

Most users of drugs have a psychological weakness. They are usually people without a future, but it could also be people with a lot of money, who want to perform better. They are in any case, people who are not in balance with their environment and within themselves. This can be prevented by no longer permitting financial excesses and to ensure that every person in the community has a bright and sunny future ahead. At the same time, you could take the wind out the sails of the criminals, a knife that cuts both ways. Or do we want to continue fighting against something that we cannot beat?

Poppy fields in Afghanistan

I just saw on the Dutch news that there are plans to destroy the poppy harvest in Afghanistan. The Netherlands are against it, because it causes the farmers to oppose the Dutch military[8]. And what about our kids? The Taliban earns a lot of money on the poppy harvest and then buys weapons against our troops. If we agree to pull out our troops and then burn the poppy fields, our soldiers are no longer at risk and take away the income for the Taliban. Looking for simple solutions!

2.17 Gambling

Gambling has already destroyed many lives. Gambling addiction is a serious problem in our society and stays largely within the criminal sphere. It is also widely used for money laundering. Gambling is almost everywhere on a large scale: gambling on horse racing, football, tennis matches, in casinos, and all kinds of lotteries. There is even a city in the United States of America, Las Vegas, which runs entirely on gambling. There are people that gamble away hundreds or thousands of dollars in a day. Another example is that gambling on sporting events often leads to bribery scandals, as in the German football league in early 2005 and later that same year in the Belgian football league.

But this is nothing compared to all those families, who have a family member who is gambling all their money away and putting their lives at risk. Often the gamblers fall into the hands of criminals and are threatened by them. Often they are forced to regularly carry out criminal activities to pay back the money they have lost.

What about the stock market? You are not allowed to conduct trading in shares on the basis of prior information. But without (prior) knowledge, it is just gambling! It is astonishing to note that this stock exchange plays an important role in our society.

It is absolutely untrue that people act without prior knowledge, because

8 RTL-journal in February 2008, the Netherlands.

information is a volatile and somewhat invisible asset. There is no smell or fingerprint to it. The probability of tracing the use of prior knowledge is minimal. The opportunities to gain an enormous amount of money are huge. The pressure on securities dealers is high. So you can imagine that when they have prior knowledge, they simply make the deal without telling a soul.

2.18 Conclusions

I can go on and on telling you what is wrong with our society. But there will always be issues that we have to deal with in our society. It is difficult to make everyone in a group of people happy. This should, however, be what we are striving for. Because now there are an awful lot of dreadful issues in our society and some of which are threatening the survival of humans and animals on Earth, it is time for drastic steps. The happiness and health of many people are at stake.

I would like to draw three important conclusions on the basis of the previous information:

1. the problems are partly caused by the nature of man, which can be explained by the evolution of the human race. This means that it is also necessary to deal with the people themselves by educating and teaching them to become 'better' or 'different' to fit into the new society. From the point of view of the evolution the need to seek power and domination can be stated as primitive needs. That is not going to change by itself.
   Children are not small adults, but must be educated, to be able to function as a balanced adult when they have grown up. As a child young boys learn to fight and play with guns. Nowadays children are watching violent TV-programs from an early age. This is reinforced by aggressive computer games and what is shown on the internet.
   Of course man wants to move forward, and we should develop ourselves. That is why we are here on Earth, but that does not have to be at the expense of others. Competition can be very stimulating, but then you have to do it in a positive way. In our society losing means that you also have to live in a smaller house and earn less money, in short losing in our society has a negative undertone. This ends in lots of traumatized people, who avoid the competition or worse still, are opposing and undermining the system;

2. on the other hand <u>the problems are the result of the systems that we use</u>. Capitalism is the institutionalization of the selfishness, being one of the most fundamental, but less pleasant qualities of humans. This system is based on competition (the strongest wins), creates losers in society and accepts this as normal. That is asking for trouble.
   Communism on the contrary is no good for the individual. There is little attention for the psychological aspects of man. The introduction of such a system is therefore supported by a huge government administration to suppress any form of opposition. The collective is the most important and people lead a life of anonymous uniformity. But in fact, the distribution of wealth under a communist regime is no different to a capitalist system, with all its disadvantages and problems.
   This means that we must migrate to a more advanced system that can focus on the psychological and social aspects of man and society, to ensure fewer problems and offer more for the future. However, this should be a system that contains a total concept and not one that is only solving part of society's problems. Only a total solution can win sufficient support from the world population and has a chance of success;
3. <u>we are lacking a vision on society and knowledge and understanding of the total universe in which we live</u>. This causes society as a whole and many individual people to be unguided projectiles. We will have to guide science onto another track, and through education make people aware of the situation we are in. The output is still the result of the inputs and processes that take place between them. A familiar saying goes: "garbage in, garbage out!". The more we take into our hands regarding the upbringing and education of our children, the less impact we have on the final result. A broad-based vision is therefore a prerequisite for a society to be able to build and lead the way.

What can we learn from the above? An important lesson could be that we have to move from a reactive society to a more proactive society. A society characterized by a shared vision, ensuring our efforts don't go in different and opposite directions, but mutually reinforcing each other, a society with a vision for the future, which is also working in the same direction, instead of a society that degrades the Earth as if it was firewood.

Educating people
There are a number of things by which every human being is formed, such as:

- education (at home and in school);
- hereditary properties (physical, but also natural character);
- environment (with whom do you mingle);
- events (positive or negative).

We focus less and less attention on education. The education should not cost too much. You cannot change your hereditary characteristics, but you will be punished for it, if you do not have special qualities: your chances of a flourishing future are a lot smaller. With whom you mingle is for a large part determined by your birth, and you cannot change this. And on the events you also have very limited influence, because sickness or having an accident, in principle can happen to anyone. And even illness or an accident of a relative can significantly affect your own life.

With the above in mind, we should be able to change some things:

1 to give more attention to education (at home and in school);
2 to give people equal opportunities regardless of their genetic properties;
3 to lift the environment as a whole to a higher level, ensuring children do not come into a criminal environment;
4 to treat people equally on negative aspects and give more opportunities on positive aspects.

Imagine what an impact this can have on educating people!

Insight and knowledge
I would like to name a remarkably positive point: more and more people are starting to realize that things must change. But most people see no escape from our current society, as we know it. They accept the system as a fact and try to work towards improvement from within. Many people end up with things like religion, spirituality and the paranormal. How many people nowadays are watching TV-programs such as Char and Derek Ogilvie, meditate or are studying Buddhism? More and more people are studying and learning about the spiritual world. In addition the religious groups can rejoice in a growing interest. In that respect it also seems as if we are

all looking up at the sky in hope and gradually realizing that we shouldn't expect help from anywhere else.

Make a choice

On one hand you can see this as an escape from everyday reality, but on the other hand, you can also see this as a gradual acceptance of the fact that there is more than the physical reality. I think that both observations are true. The second is the more positive interpretation. But we can do with both. The more people come to this insight, the easier it will be to take the next step as a society, because people need a picture of a better society and also the confidence that we can get there together. And that is exactly what I am trying to achieve with this book: to show everybody that we have a choice, that there is an alternative to this world in which we live and that we can get there.

Take responsibility

To have a choice also implies a responsibility, namely responsibility for the outcome of the choice. That is what we have now and that we have for the future. To say that we have no choice, is actually trying to deny that we have a responsibility. But believe me; we are all responsible for this world. It is no use trying to deny that. Therefore make a positive choice for a new society. Take your responsibility and you will you feel better than ever before.

Chapter 3

If we continue the way we do

3.1 Introduction

What if we just continue with this society? This is obviously a question that many people will ask. Why would we turn our way of life upside down and run the risk of a new way of life not working? The answer is simple: we will always be forced to change, anyway. The evolution has shown that many species have come and many have disappeared. Civilizations have come and gone or merged into others. There will be diseases and natural disasters taking place, which will affect our way of life in any case. But the most important and most certain reason, I think, is because we are consuming all the natural resources on Earth at high speed. A system based on growth cannot continue on a planet of limited size. While realizing this, the future becomes brighter. In this chapter, I will outline the most likely possible scenarios.

3.2 Change is coming

Where are we now in our society? If I look at the life and welfare in the Netherlands and indeed the entire capitalist world, I see it as a parabola with the point upwards, "our society, in its current form, has passed its peak". See scheme 1 on the next page.

I believe we are somewhere at the top, about where the arrow points to. How do I come to this conclusion? The previous chapters have already made this clear to a certain extent. During the development of human mankind the conditions of living were getting better and better. We were becoming more prosperous, life was becoming healthier, happier and the average age at which people die is still rising. The environment never had any problems and there were always enough natural resources. The sky was the limit!

But now the situation is changing. Let us take the Netherlands again as an example: the environment and the world population are increasingly imposing restrictions on us. We cannot continue just investing in more prosperity, because we have to start investing in the environment. This will be at the expense of a piece of prosperity. The longer we try to go on with our current system of growth, the harder we will go down. I am not saying that the way down is inevitable, but in order to escape this downfall we need to change and to do that very quickly, because the longer we wait, the greater the dip will be.

In addition, we need to compete against low-wage countries, putting the wages under pressure not only in bad times, but even in good times. Entire industries are disappearing to low-wage countries. Again, this puts pressure on our prosperity.

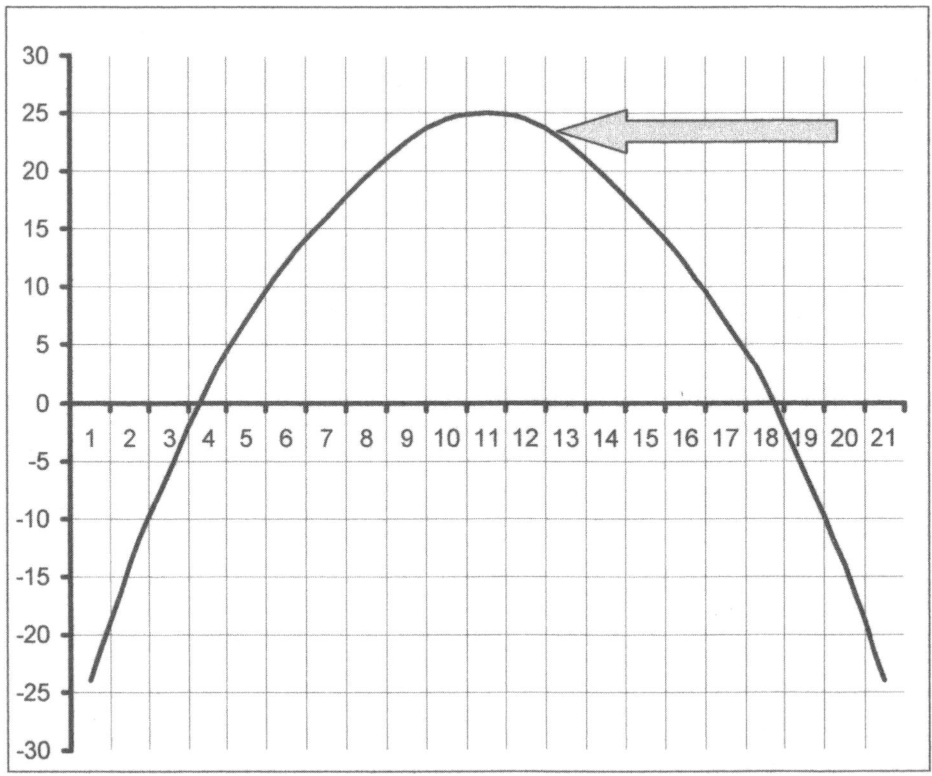

Scheme 1. The direction in which the capitalist society is developing

The only thing we really have left is the service business, but for how long? The conditions of employment and the working conditions are getting less. Private offices used to be adapted to the individual employee. Today, there are large office floors with often hundreds of standardized flex office spaces and if you're lucky, there is a space free for you, because there are less office spaces than there are employees. This saves the company a lot of money. And no whining about too much noise or too little privacy, because then you can leave. We still had good pensions and early retirement schemes a number of years ago, but this situation also has changed. The early-retirement schemes have been abolished and if stock prices fall, your pension is no longer guaranteed and there is nothing you can do about it.

Moreover, the growing world population and emerging economies will put pressure on the world resources, causing us to pay much more for various products such as petrol, gas, electricity and clean drinking water.

Finally, many companies have been taken over by foreign investors. When they leave, where should we go to work? The only thing the government can do is to keep taxes low for companies, but the government must also have money to work with. The only thing they can do is raise taxes for the ordinary citizen. Again, this will be at the cost of our prosperity.

In short, we are over the top.

We strive for an everlasting higher level of prosperity and are prepared to invest almost all our time and energy for it. This will gradually come at the expense of welfare. It is time for something new. Eternal growth is not the solution.

Perhaps that this parabola is typical for all life on Earth: you are born and are becoming taller, stronger and smarter, until you reach the top. One person may be able to maintain his or her top a little longer than the other, but in the end everyone has to go down, until we die. The same goes for the economy and civilizations. The big difference is that these last two parabolas each consist of a row: ongoing wave movements. And perhaps our lives are no different, but we do not see it like that yet.

3.3 Scenarios of the future

The most probable are the next three scenarios of what may happen in the future:

A. **Controlled change**: allowing the economy to grow in a controlled manner by making mutual agreements and by preventing major environmental disasters.

B. **Forced change**: global agreements about tuning the economy on the environment and the use of resources have failed or are not effective enough. This will inevitably lead to climate change and environmental and natural disasters which will drive us in a certain direction.

C. **World War III**: just like as in scenario B we do not succeed in making agreements about the raw materials. Wars like in Afghanistan and Iraq could lead to a world war for raw materials and for world power with devastating consequences.

3.3.1 Scenario A: controlled change

Controlled change is a scenario where we try to bring the developments in reasonable equilibrium with the environment. This is a variant which we are now working towards in our world. We try to make deals, by which we spare the environment so that no major changes in the environment will take place. Have you already seen in nature documentaries how various ecosystems on Earth are working together? Have you ever wondered how food chains are working? These are very complex systems, which are the result of millions of years of evolution. In a period of about 200 years, we have brought danger to almost all ecosystems, and hence our own survival. Thus, it becomes increasingly difficult to find clean drinking water on Earth. The limited agreements that we have made so far will in no way be enough to change that situation. Many rigorous measures must be taken to save life on Earth, our beautiful planet as we know it.

Higher prices

The changes in the environment are a matter of large numbers. The world population has grown so fast in recent decades that all these activities must have an impact to live (or survive). To change this situation measures must be devised, expressed in large numbers. The use of too much fossil fuel should not be allowed in such a short time (in power stations, houses, cars, aircraft etc.). Cars, planes and factories should start using cleaner forms of energy. People will have to cope with higher energy prices. Houses will have to be rebuilt in order to be self-sufficient on energy (think of windmills and solar panels on the roofs). This will be regulated through money and taxes within the current capitalist system. What does this mean concretely? Very simple: all prices will go up, so that the largest portion of the population can hardly make greater use of resources. Products are becoming more expensive because many factories will have to deal with higher costs as a result of making their production process cleaner and less energy consuming. The use of fossil fuels will be put under pressure by additional taxes. From this tax-money compensatory measures can be taken. People will have less money as a result of previously mentioned investments, which will cause the domestic economies to go down. And all this under the motto: everyone must contribute to a cleaner environment! But is that really the case?

The party continues

The capital of the rich and powerful has grown enormously in recent years by profiting from the securities trade and the use of cheap labor in low-wage countries. They do not worry about all these measures. The costs of placing solar panels and windmills in their huge gardens are negligible compared to their total capital. They can keep racing around in their petrol slurping Porsche's and Ferrari's (or 10-cylinder hybrid Lexus limousines) and flying to a friend or family member in an afternoon 500 kilometers from where they live and back or even go shopping in a metropolis. Then they give a nice Christmas donation, a capital for poor people, but this is peanuts to them. Meanwhile, their capital cheerfully keeps on growing, because most of them do not pay taxes. They even get money back for their Christmas donation. But not for the simple people, because they still need to go to work, because otherwise they cannot make it till the end of the month. All their money is being used for survival. These are the people who pay for the rescue of humanity and have nothing at the end for themselves. If they have bad luck some of them get fired or sick. For a lot of them their future is very uncertain.

Is this fair? Is it social? Is this the equal sharing of joys and burdens? Furthermore, all these measures do not fit in with the liberal capitalist system. People in the Netherlands are asking questions about the electronic toll highway system, because everyone knows that the rich will keep on driving in the same way they always have. It is only the poor people that will disappear from the highways in expensive times. But this is just the beginning. What if the people in poor countries are no longer allowed to cut trees to make farmland? Where are they supposed to live? What freedom will be left? The rich and powerful will determine what can and cannot be done. You will have to work hard until you die for the wealthy people. You do not even have time to educate your children well.

Tight control

Of course you think that everything will work out differently and that there will be other solutions. Well, let us assume that we succeed in a large scale development of alternative energy, energy that is not polluting the environment and of which sufficient can be produced at a reasonable price for everyone.

In order to keep the dream alive so that everyone can have a beautiful and rich life, we stick to the capitalist system with its economical growth. This means that the world will continue to grow and consume, because we want more: more sales, more production, more profit and more workers.

However, not everyone can live in the same beautiful place and not everyone can go to the Bahamas on a vacation, because they are simply not big enough. Moreover, because more and more people are coming, there will be less and less nice, quiet places. The prices will be even higher. This will lead to an increasingly severe struggle for luxurious things. The criminals will also keep on playing their game even stronger than before, because we are still using money. If we are going to lose this fight, then this can lead to total anarchy and chaos. The power will then be exercised by brute force, just as in the thirties, in the time of the Mafia. That is what we want to prevent at any time. Therefore the measures to keep people under control will strongly increase. People will be followed by cameras, GPS systems[1]and tapping phones. The checking of all electronic traffic will be a continuous activity. In some cities there are already so many cameras, that people are filmed several times a day. If the intelligence office has the presumption that you do things that may be threatening to the system, they will do everything in their power to make you harmless. This does not mean that you will be shot dead or put in jail, but they can deprive you of all your opportunities to build a life in this society. Are we talking about a free society?

Well, suppose you lead a neat life and work hard, aren't things going to improve for everyone? Wake up! This is the capitalist system! That means the law of the jungle! Competitors you need to outsmart and ensure that you make a lot of profit and do not go bankrupt. If your competitor can offer lower prices, you will not sell anything. You will therefore have to be cheap. That means at low costs, keep wages low and not give away anything, unless you get some-

1 GPS stands for Global Positioning System. A system that makes use of satellites orbiting around the Earth. These satellites can provide the exact position of an object on Earth.

thing in return. You can go sharing things, but then you will never get rich and people will be able to tell you are a communist. That is not the intention. One solution is to buy your competitors or make cartel agreements. In both cases you are killing the system and that is not allowed. Yet it happens and it will only get worse. Globalization is the latest step towards total domination for a limited group of multinationals with huge amounts of capital.

Growing world population

Suppose that the world population grows to ten billion people, and one billion of them are either a millionaire or billionaire. One billion millionaires and billionaires, that is incredible! They all want to live in big houses on large pieces of land, drive in large cars and make distant voyages. The quantities of raw materials on Earth are limited. Where are those other nine billion people going? They are working for the rich to make their lives beautiful and easy. And if the rich want to share a bit, someone else may also get something to enjoy.

Let us assume that three billion people will actually succeed in getting jobs close to the wealthy and rich people, and then we still have six billion people left for whom there is hardly any room. They have to work hard their whole life for a very small wage (because there will hardly be any pensions left). We are forced to work longer by the government. Stock Exchanges are unreliable as a foundation for retirement. You have to save money yourselves for your old age. There remain three groups of people: a relatively small group of wealthy people, a very large group of poor people who have little access to resources, and a group of people in the middle who manage the large poor group. The people from the middle group are assistants of the rich. And because the poor also want something in their life, the problems with unfair people, criminals, illegal immigrants, refugees, violence and war are much greater than they are today. The one billion millionaires and billionaires will not worry about this because at that time they will all live in neighborhoods surrounded by large walls, which are well guarded by security guards patrolling the area on a regular basis. The assistants of the rich ensure that the money continues to flow in the right direction. And if they get the chance, they will also attempt to become rich. This will often occur as a result of meeting the right people, doing secret business or by putting pressure on other people. But that is not their concern.

And what about health care? What part of the ten billion people will be allowed to make use of the health care? Certainly the one billion rich, but what about the rest? And to stay healthy, you will need to eat healthy food, but as more and more people are coming, healthy food will become scarce and expensive. Therefore this will not be available for the ordinary people. Altogether they will live shorter and have less healthy lives.

Dictatorship

Six billion poor people cannot be controlled in any way. That is simply too many people, and each of them can think by themselves. A lot of them will look for freedom and a means to live. The rich people in the capitalist system are a bad example for others by taking what can be taken. The poor people will follow this example. In the poor countries people also want to build a life. They will continue logging and burning rainforests, because otherwise they cannot use agricultural land and have no form of livelihood. Many plant and animal species will become extinct and there is even a chance that the overall ecosystem of the Earth will be disturbed. To prevent this, many countries will form a dictatorship, such as China. Massive resistance will occur, which will end the dictatorship in the long run. But at what cost? We have seen that such dictatorships are able to maintain themselves for many decades. The Earth might still be livable, but from a social and psychological point of view, there is a non livable situation. A sustainable environment alone is not enough for a happy and free life.

If we can only save the world by introducing so many rules and/or a dictatorship-like way of control, you can ask yourself for whom we are trying to save the world, as we know it? All the new rules will of course be expressed in cash, and if we do that they will only have an impact on 'ordinary' people. There will never be equal opportunities and possibilities in this way. The oppressed majority has no life, although they are not dead. In this way we are maintaining something in the benefit of a small wealthy minority, who do their own thing. That really looks like the old communist dictatorship! When it comes to this, you can even ask yourself whether we are better off if we give up the Earth, as we know it, as a place to live, if life for most people is not really life anymore.

Scarcity of raw materials

In short, because of the continuous growth of the world population, there will be a shortage of all kinds of raw materials, which are only affordable by the people with big pockets of money. Resistance will occur, causing all kinds of conflicts. In addition, a large poor population and a disruption of the ecosystem will increase the chance of an outbreak and rapid spread of epidemics, causing death among a large part of the world population. Of course, the wealthy people also run the risk of becoming sick and their way of life is also in danger. These people are not in the production. When a lot of people get sick and die, the production of all sorts of products will come to a standstill. It is possible that this will lead to a new world order, where there are a couple of billion people left, who will start a new form of society. As a result of the massive pollution of the Earth and depletion of raw materials, this new society is necessarily more focused on survival than it is today.

So even if we succeed in managing everything reasonably well (please note: compulsive through rules and laws), then today's social systems will lead to a growing world population, which will bring uncontrollable problems in the long term. All the solutions that we think of are always temporary. It is only symptom control. We never meet the real cause of all problems. Effectively addressing the problem may be postponed for a while. Some problems even disappear by themselves after a short period. But this is a very fundamental problem. If it is really not going to be addressed, it will consequently cause the collapse of a very large part of the world population, if not the entire world population.

3.3.2 Scenario B: forced change

Another scenario is similar to the previous scenario in the beginning, but that's not so strange, because we depart from the current situation and maintain the capitalist system, the foundation of our society. The big difference lies in the fact that we are not succeeding in making agreements (on time) on a world level to prevent drastic changes in the environment. What does this mean?

Increase of CO_2 and temperature rise

In the United States and China there has been coal in the ground for hundreds of years. However, if we are going to use this on a large-scale for extracting for example diesel fuel, the CO_2 content in the air will rise in record pace. Other toxic substances such as sulfur dioxide will increasingly enter the atmosphere within a few decades. There are already hundreds of billions of tons of methane, which are released from the thawing permafrost soil on the mainland around the North Pole by the warming of the Earth.

In combination with global deforestation, as a result of the forestry and the establishing of fertile farmland, there will be on the one hand an enormous increase in CO_2 and on the other hand a reduction of CO_2-absorbing and oxygen-producing forests. But the forests do not only have a function when it comes to the regulation of CO_2 and oxygen, but also for the cooling of the Earth. Where there are lots of forests, there is also a regular rain fall. The rain has a cooling and purifying effect. Where forests disappear, there are green plains at first, but in the end, due to lack of rainfall, only deserts remain, where temperatures are much higher.

The North and South Pole, with all their snow and ice, are also contributing to the cooling of the atmosphere. What if all these large cooling systems are gone? The end result will be a strong increase in the overall average temperature on Earth. This might subsequently cause the seawater temperature to rise, causing the seawater level to rise also all over the world. The question remains whether this will be worse than the shortage of fresh oxygen that could arise in the course of time. The increasing number of people is heavily dependent on sufficient oxygen and most of them are living in the vicinity of the seas around the world.

Nuclear energy

Another source of energy which is becoming increasingly popular, as a result of rising oil prices, is nuclear energy. People are very worried about nuclear energy especially about the nuclear waste. However, this is relatively small in proportion to the damage that a nuclear power plant can cause on a daily basis. Search for 'Sellafield' on the internet and read what impact such a power plant can have on the living environment of people. An accident with such a power plant is not in proportion to its benefits. Search for "Chernobyl" and read what the effects are on life in that area where the nuclear power plant exploded now more than 20 years ago. Because the area is in fact non livable, the people there are left to their fate. It is not acceptable to handle people in this way! Despite the fact that this is the only major accident involving a nuclear power plant so far, it cannot be justified to take such risks. If something goes wrong, the lives of the people who made the decisions to build such a plant are not at risk. Then it is easy to choose.

Impending disasters

It does not matter what kind of disaster(s) we assume. If the environment changes due to higher CO_2 levels, deforestation, eradication of plants and animals, this will affect a large part of the world: the ordinary man in the street. The rich and powerful will do everything to maintain their present way of life. They have houses all over the world and with their Private Jets they can fly anywhere. Nothing within the current system can stop them. Because they have power in this world and thus bear responsibility for the policies, they are also responsible for the result. It would therefore be reasonable to judge them for this result. The likelihood is however that the people will ask the politicians for an explanation, because they have always been in the picture when it comes to policy. Behind the scenes, it has been the rich and powerful, who have pulled the strings of the politicians (in the United States multinationals have lobbyists in Washington, using lots of money, securing the interests of the multinationals through the politicians). Hence, they are really only puppets in the big picture.

According to the current calculations (in 2008), the production of CO_2 from Europe and Russia will rise only slightly over the next 15 years, a little more in the United States and a lot in countries like India and China, countries that are recently participating in the world economy. Given the fact that Europe has been barely able to curb their CO_2 production, despite all sorts of rules and laws, we can safely say that the CO_2 content in the coming years will adopt dramatic values. Especially if we also take into account the further decreasing absorption capacity of the Earth of CO_2 as a result of the destruction of most of the forests on Earth.

Total disturbance of the Earth as a coherent ecosystem

It starts with average temperatures rising slightly, irregular rainfall (long periods of drought and short periods with lots of rain), and storms in periods and places where they previously did not occur. Then the change of weather phenomena turns into natural disasters: floods, dam breakthroughs, landslides etc...

The major natural disasters will then lead to landslides in the earthly human relationships. Entire countries will disappear. Complete populations, hundreds of millions of people, perhaps even billions, will have to flee to other countries and must leave their homes. People will literally go and jostle each other. Countries will be flooded with refugees. There will be many millions of deaths. This can all be avoided if we take measures in time.

The strong (richest, most powerful, smartest, most flexible and most healthy people) in society will probably be able to save themselves (*It is still unclear what role genetic manipulation in this context will play, but in the longer term this could cause the phenomenon of 'the strongest' in all respects to be concentrated among a small group of people*).

The total disruption of the ecosystem, reducing the plants and trees, the release of toxic gases by the thawing of the polar caps, it is even conceivable that the Earth's atmosphere is becoming non livable to humans, because there is not enough clean air.

In any case, an extreme disruption of the ecosystem will lead to extreme reactions of the Earth, which we are unable to stop when it occurs. We will have to wait for tens, perhaps hundreds of years until the Earth itself has built up a new balance, in which we as humans are able to build a life, without having to be punished all the time by the Earth.

Then the whole game starts again. But the step back that we have taken then becomes larger than ever before. There will be a lot of rebuilding to do, but the rich and powerful are still pulling the strings. Actually, we have learned nothing and are heading off to the next disaster, which we again will cause ourselves. There is a large possibility that conflicts and wars will break out worsening the situation. After many years, there will be a new equilibrium emerging which is the result of natural forces and the law of the strongest, all this at the expense of many lives. How much further will we be then in our development in comparison to the Neanderthals?

This is an even more negative scenario than the previous one, because it is debatable whether there is still room for mankind, when a series of large environmental and/or natural disasters have occurred and the existing natural balance is disturbed. The disturbance can be so serious that this leads to the extinction of mankind or a minimization at least, because we have not altered the situation ourselves, we can only wait and see whether the new situation still offers opportunities for humans to survive.

3.3.3 Scenario C: World War III

Another scenario is also quite possible, the emergence of the third world war. Everyone expects that if a lack of raw materials arises, this could lead to increasing conflicts. Because until now there has been growing prosperity everywhere, some scientists are less pessimistic. But what they have overlooked is the fact that as a result of the shifting of power the rich countries could very well be confronted with a declining prosperity.

China and Russia

A country like China is quite capable of producing virtually all the necessary products in their own country. Countries like the United States and the EU can only compete by offering cheaper products, with support in the form of high level knowledge or special products in niche markets[2]. To be able to offer products cheaper than China, will be almost impossible, given the high cost of living in the West. The other services and products will offer employment to a very limited group of people. The result is that the prosperity in the western countries will decrease over time. Many young and educated people will depart from Europe and the United States. These areas will become decayed civilizations. Lost glory, worse than in the former Soviet Union. The Soviet states have fallen apart and are greatly destabilized. Russia knows how to sustain itself as a world power, thanks to the huge oil and gas reserves. The combination of Russia - China is a dangerous one: China as an economic and military world power and Russia as an energy supplier of the economic giant China. The western countries obtain their energy for a large part of the Middle East. It is not clear for how long they can or will act as an energy supplier. What if the deficit of raw materials happens faster than expected? What are the countries without natural resources going to do? The United States are almost at the end of their period, but still have an awesome army. The United States are a democracy and China and Russia communist dictatorships. In time, this will inevitably lead to a clash. China is becoming the most powerful country in the world in all respects.

---

2 'a specific, defined part of a market. Niche markets have less competition than the main market; some niche markets have even one provider.' From the English Wikipedia-site.

The Middle East
Another source of escalation might be the Middle East. The Islamic countries are increasingly at odds with Western countries. The areas of the Muslim countries are of interest to the West as a result of the present resources. It is very well possible that this will lead to more serious and severe wars than in Afghanistan and Iraq. Once the actions of the United States and NATO countries are no longer tolerated by countries like Russia and China, things might go completely wrong. If these other countries start to mingle in the conflicts in the Middle East by supporting the counterparty(s), such a conflict might escalate to world level.

The Arctic
And what about the North Pole? Here, the United States and Russia meet once again. They are now waiting for the Arctic to melt, which will not last for many more years, unfortunately. A race to the best oil drilling places will occur. If we cannot agree exactly which part is from whom or one has the idea that the other has achieved an essential advantage, then this might lead to an armed conflict.

Defense
There are several places where a third world war might occur. The fact is that the EU countries, such as the Netherlands, have invested very little in defense in recent years. They will not be able to resist a strong force from the east (north, middle or far east). This will be able to penetrate directly to the Atlantic Ocean. It is debatable whether this is necessary with regard to the Chinese, because they are taking over more and more companies, which can have a strong corrosive effect on Western culture and way of life in the long term. We will be forced to do whatever they want or else we will have no work and no money to live off. If, it however, comes to an armed conflict, then we cannot possibly stop them. We are also, by the great distance and the vastness of China, unable to retaliate adequately and effectively. If we do not quickly surrender Western Europe will look like the ruins of the Roman Empire in Rome. Not much more than a historical attraction of Europe will remain.

Given the events of the past, China already has a carte blanche for other countries to invade. The West has proven itself so dependent on China; it will not take action if China attacks other countries like Taiwan who will fall within. This is what we have already seen in Tibet in the fifties.

In principle, we exclude such a conflict, but isn't the history of the world a succession of wars? Why do we think that a third world war will not take place, or do we still think it is possible? Why do we not build a strong defense against a potential aggressor? We believe that we are able to prevent any conflict with enough communication and intelligence? How can we expect this if we ourselves are living by a system based on the law of the jungle, not fair and oppressing against people? An intelligent outsider will see this too and not be fooled by our slick talk. I think we are overestimating ourselves and in this way are putting our way of life at risk. Once again, the rich and powerful think they will not be harmed, because they have money and houses throughout the world, where they can elope in an emergency situation or in case of an armed conflict. They live without borders in fact. But do not think that they will also take care of you. Certainly not when they are far away and have arrived safely!

3.5 End conclusions from the scenario's

In conclusion from the foregoing the following conclusions can be made:

1. **the first scenario, controlled change** will lead to increased regulation, which will ultimately be experienced by workers as class oppression, dictatorship. The population will be more clearly divided in two: a small rich part and a large poor part. As we continue to build on the current capitalist system, the continued growth of the world's population will in time lead to problems with the ecosystem on Earth;
2. **the second scenario: forced change**, the outcome will be determined by the outcome of climate change. This will cost a lot of people, if not all, their lives. A new balance will occur. It is difficult to predict what the new position of man will be and at what level of prosperity;
3. **the third scenario: World War III** may arise as a result of shifting power. Parties will try to make use of this, others will feel threatened. A conflict could suddenly arise. This will cost many lives and the outcome could mean less prosperity and freedom for many people.

On the basis of these conclusions, I think we can say that all three scenarios do not sketch a rosy picture of the near future. Wars, environmental or natural disasters are not things we are hoping for. These seem inevitable because of the growing world population and the decline of natural resources. The alternative is increasing regulation and rising prices, causing to limit the freedom of many people. In the worst case, this will result in tight dictatorships. A scenario which on the one hand is the result of adherence to the idea of growth, and on the other hand the wish to maintain the current contrast between the rich and poor. We will therefore need to think carefully about which way we want to go. Life is not just about having pleasure. Let us think again about the world we leave to our children and grandchildren. They will have to go on in this world. Do we still want to offer them a bright future and a world to live in?

This not only means following scientific research, but also looking around you, doing some logical thinking and finally letting your heart do the talking. And believe me, you really are not the only one. Just take a look on the Internet: various sites are talking about the same problems, not only on private sites, but also on business sites.

I talk to a lot of people about the situation in the world and they all agree that only a drastic turn can improve matters. They know that what we are doing now, offers no definitive solutions. Only a major change can offer a long term future for all people on Earth, poor and rich countries.

Of course it's easy to create a negative image, you will say. Enough criticasters. OK, maybe so, but do you want to suggest there is no reason to be negative about the future? How long do you want to keep convincing yourself that there is nothing wrong? Until it is too late? Who knows, maybe we have already passed the turning point and what we have done is no longer reversible.

The reason why I have outlined these scenarios is because I believe that there is a positive scenario.

Chapter 4

Are there already positive developments?

4.1 Introduction

The people who are fifty years or older today are the happiest people in the Netherlands, according to the statistics. Only half of them are still working. Many of those early retirees have had good retirement payments or early retirement arrangements which has enabled them to enjoy their old age. Some of them do small jobs, which altogether makes for a fantastic overall income. Very nice, but this is unreal when you realize that the government has just decided that the current generation of workers should go on till they are at least 65 years of age, because otherwise all these pensions would become priceless! [1]
And the current fifty-plus generation is also traveling a lot, all around the whole world. I heard a story about a couple that traveled through the United States for three months in a luxurious camper. They drove about eighteen thousand kilometers. Do you know what the fuel consumption of a motor home is? One liter for three or four kilometers! In three months they had burned just as much fuel as an average Dutchman does in more than three years! And you think we are going in the right direction? This is just one example, but it illustrates that many people only think of this moment and how they can have the most fun without thinking of tomorrow and about being a good example.

Many people, who are satisfied with today's society (people with a 'rich' life), will say that we are already doing well. That we are already doing everything we can to help the poor and unfortunate people and to save the environment. That everyone has equal opportunities in this society, and that there is no other way. But is this really true? Are we already doing the right thing and is it even possible to do well within the limits of the current system?

An important principle of love is to give and not expect anything in return. That is the big difference with the capitalist system; against everything you

1 In June 2008, the committee Bakker suggested to the Dutch government that workers should continue until their 67-th year of life before they can retire.

give there is always something to receive. How can we convince people to do something for which they do not immediately get something in return, such as doing something good for the environment?

Are there any good examples of progress in the right direction? Of course there are. First, you have to take a good look and secondly, the examples you find need to be reviewed from a positive perspective. Let us look at some recent developments.

4.2 Recent changes

Social Responsible Enterprising

We are working on the environment by giving our products a green character: green electricity, energy labels, green shares etc. Environmental responsible enterprising is strongly connected to the term People-Planet-Profit.

People Planet Profit

"People Planet Profit (also called the three P's) is a term of sustainable development. It represents the three elements, people (all the people in the world), planet (our planet with its environment) and profit (revenue / profit), which should be combined in a harmonious manner. The term was invented by John Elkington, a consultant in the field of sustainable development.

If the combination is not harmonious, the other elements will suffer. For example, when too much profit is being prioritized, then the people and the environment will suffer with such things as poor working conditions or destruction of nature. Conversely, the slogan also features the profit as an essential part of development that should not be neglected".[2]

Regarding People-Planet-Profit, there is even a Dutch website: www.peopleplantprofit.be that is supported by a number of major companies, including Rabobank, Triodos Bank, SNS Reaal, PGGM, and Delta Lloyd. On this site positive initiatives within the framework of social responsible enterprising are regularly presented. Let me mention a few.

---

2 Quote of the Dutch Wikipedia site.

Examples
There is much attention for the extraction of fuel from algae. Algae are grown in salt water. The sea covers the largest part of the Earth. This means that the potential to grow algae is huge. Biodiesel can be extracted for algae, which can make planes fly.

A company in Barneveld, Paper Foam, produces packaging from potato starch. This replaces the plastic packaging of, for example, CDs and is 100% biodegradable.

All kinds of paint may no longer be produced on a chemical basis, water based paint is to be used instead. Or paint mixed with solar cells applied on steel profiles of office buildings. This can be used to generate electricity, which could lead to huge energy savings, if widely applied.

Or maybe think of power plants on top of waste dumps? They produce a lot of methane gas, which can be used to generate energy. This is already done in Sao Paulo, Brazil.

We are trying to reduce the CO2 emissions from cars, inventing hybrid cars and are trying to make cars which run on alternative fuels.

Many positive actions in modern society are initiated by individuals and small groups, people who work for the welfare of others; because they believe that this is the correct thing to do. I am talking about ordinary people who come to great deeds. I can remember a case about a boy in the slums of South America, he was born there. As a child he hardly played with his friends, like other children, but he helped the elderly people with whatever he could do. More and more people started to notice this and after many years, this led to the creation of a shelter for the elderly in his neighborhood of which he became the director. Very extraordinary, but it does indicate that wonders are still possible.

But I also want to name some initiatives from a number of rich and powerful people, because every contribution is worthwhile. Like Microsoft founder Bill Gates, who together with his wife, founded the Bill & Melinda Gates Foundation in which they put a lot of their private capital to establish projects in developing countries in the field of health and education.

In 1982 former U.S. President Jimmy Carter and his wife founded "The Carter Center" that is committed to human rights and is combating unnecessary human suffering, especially in poor countries around the world.

Swedish millionaire Johan Eliasch has bought approximately 160,000 square kilometers of Amazon rainforest in order to protect the ancient forests which he sees as the lungs of the Earth and the producer of fresh water. During a meeting of businessmen in July 2006, he proposed that the entire Amazon rainforest should be bought[3].

In California, many initiatives are put into practice against pollution of which the rest of the world has little or no knowledge. For example, think of small windmills on roofs of houses with a flat roof that can be attached to the walls to use the rising air to make electricity. In combination with solar cells houses can be self sufficient in producing energy using these techniques. Even a country like China has done a lot for reforestation in recent decades, if only to ensure that they would be completely dependent on imports of timber from other countries. And there are many countries in the world, where people have started to compensate the loss of forests by planting new forests. The repent can also be seen as a positive element.

Then there is the World Social Forum, as a counterpart to the World Economic Forum, the WTO and the G8-meetings. The proposals of the WSF have the objective of a globalization including solidarity, respecting human rights, respecting all the people in all the nations, both men and women, respecting the environment and are based on social correctness and equality. The WSF is against the dominance of capital and any form of imperialism.

We can go on and on naming examples. We therefore may conclude that we are already trying to turn the economy in favor of the environment. These results still come back to the realization that something is wrong and that we obviously need to do something. It is nice to see that people are trying to do something for the environment within the scope of the current economic system. People are aware that environmentally conscious business is good for the image and can even be profitable. The group of people who think about what they do here on Earth is increasing. And it is precisely that change in the

3 According to an article on www.oneworld.nl .

way of thinking that could provide the sparkle to the next step in the development of our society.

On the other hand, these are just drops in the ocean, because the really big conventions such as Kyoto, which discussed amongst others, the heavy industry, are not making a lot of progress. People are afraid to commit themselves to investments in favor of the environment and at the same time running the risk of losing the competitive advantage and as a consequence the support of the shareholders.

The people in Europe are far ahead in their thinking about the environment. The people in the United States, the world's largest producer of CO_2, are also becoming more aware thanks to the movie of Al Gore on CO_2 emissions[4]. Other economic heavy weights such as China and India think differently, their economies are only just emerging, therefore the environmental pollution is growing fast. Russia is lacking behind in development by the lack of sufficient structure and coordination since the disintegration of the Soviet Union.

Europe has the chance to be an example for the rest of the world.

Growing awareness

Fortunately we are inventing more and more things to reduce energy consumption as a result of the increasing awareness, the speed of development should increase to enable us to save the environment. Perhaps this is possible, but there are still a lot of big countries which have to start to cooperate, before it is too late.

We are already starting to see that houses can supply themselves with electricity by using solar panels and windmills, and that cars can have much less fuel consumption and produce much less harmful substances, if we are willing to invest in environmental friendly engines. It is now a matter of just going on in the right direction, and with the help of government regulation ensuring that people start to invest in the future of their children and grandchildren.

---

4 Mid 2008, the leaders of the major industrialized countries (United States, France, Great Britain, Japan, Russia, Italy, Canada, and Germany) have agreed at the G8 summit to reduce CO_2 emissions by 2050 by 50%. How they plan to do this has not been discussed.

Spirituality

And finally, I want to emphasize the spiritual dimension in the world. It seems that, more and more people in wealthy countries are getting interested in the spiritual aspects of life. Some just do that because they find it interesting or because it gives them tranquility, or maybe they are looking for a higher spiritual level. Many gurus are traveling around the world to help people find what is inside themselves and to find the strength within. One guru is doing a better job than the other.

It is not yet clear whether this is a pure and truthful development, or that it is just the result of the stress of daily life and the fear of death. For example, what I see is that people and businesses start to adopt a spiritual way of relaxation, so that an even higher performance can be delivered. Or that it is used in an attempt to lift the ego to a slightly higher level than the purely physical level. People who do this just want to feel at a higher level of development than others, as a kind of status symbol. It has nothing to do with universal unification, as it is preached by the Buddhists for example.

Let us try to look for the positive aspects of things. We can say that attention for spirituality is a good development, because it makes a lot of people more loving and it gives people a more positive mindset, which can have a healing effect on the rest of the world. I expect that it will still take several generations before it has a huge impact, because many people are still committed to their 'old fashioned' religion[5] and are too busy with the non-relevant issues of their religion. I mean things such as the compulsory wearing of headscarves, black stockings, skirts, the mandatory fasting, not being allowed to watch TV, etcetera. This is not what it is all about. The spiritual impact is limited to the essence and that is a good thing.

There are more and more people who believe that all spiritual beings in the universe are connected to each other and thus being the ultimate means of transportation for development and deepening. This goes beyond their own ego, leaving behind the individualism and materialism that characterizes the capitalist system. We should consider this as a positive development, which

---

5 'Old fashioned' is between quotes, because I am not suggesting that the religion is old-fashioned, on the contrary, but I think every religion must also evolve, look forward, not stick to the past, to traditional customs and practices, but rather learn from the past, thinking about the future in a positive innovative way.

offers hope for the future. Just search on the Internet or read a spiritually oriented magazine. You will find a lot of the same ideas everywhere about the developments that are now taking place in our world, about our place in the universe and about the functioning of the universe itself. In the long term this could lead to a new way of thinking as a basis for our society. And perhaps to the spiritual society!

4.3 The Netherlands

Relatively low poverty

Productivity in the Netherlands is very high. Compared to neighboring countries, the poverty is low. Less than 10% lived below the poverty borderline in 2008. Poverty is difficult to imagine, when you have enough money and have not (yet) experienced it. That is why many people talk easily about poverty, unjustly in my opinion. Because poverty usually has an unhealthy effect on one's lifestyle, in the beginning it affects the physical constitution and after that the mental constitution. It seems no fine prospect to me, when you get in such a situation. Approximately 1.5 million people live below the poverty borderline and about half a million are just above it. The other 15 million people have a good life, especially if we compare it to the rest of the world. If we could help the other 2 million, things would be great for everyone. But do we really want this and can we succeed? Many people think of the poor people as lazy people, not prepared to work. Their situation is their own fault, they say. And of course there will surely be people who do not want to work, who do not (want to) fit into the system, but believe me, deep in their hearts they want to belong to other people and be happy like everyone else.

We can be proud that we can offer such a large proportion of the population a reasonable prosperous form of existence. In this respect most countries in the world are doing much less.

To be happy

In surveys approximately 90% of the people say that they are happy. Most people are satisfied with the work they do, despite the fact that about half of all relationships fail, many people are in the category of chronically ill patients, and centers for psychological and social assistance can hardly cope with the numbers of people that are asking for help nowadays. Apparently there is something deeper within people, which contributes to this contradiction. Who does not want to be happy? No one, I suppose? To be happy is a state of mind. You can be happy, regardless of your situation. That does not mean that you are living, by definition, in a pleasant situation. Many people think they are a victim of their situation and believe that there will be little positive change. They remain in this situation for the rest of their life and have "decided" to be ' happy' the way it is. This may not be seen by the powerful people in society as if they do not have to do anything to improve the circumstances of these people.

Landscape and Infrastructure

The establishment of the landscape and infrastructure are other positive points. This is well arranged and maintained in the Netherlands. Unnecessary risks are avoided as much as possible. When you have an accident caused by infrastructure that is not well maintained, you can appeal to the government authorities. There are also many facilities such as sports clubs, cinemas and theatres. Nature in the Netherlands is still reasonably well preserved. The Netherlands can be characterized by a great variety of different landscapes. And health care is also something that we (still) can be proud of. Until now, everyone can make use of our health care system. There are a number of doctors and dental practices in each district. In addition, there are many modern hospitals of good quality in the Netherlands across the country.

Democracy

And of course we are still a democracy in the Netherlands, where freedom is very important. Even to the extreme where we are prepared to put people's lives into danger for the benefit of the freedom of speech (think of the movie ' Fitna 'by Geert Wilders from 2008 and the movie 'Submission' by Ayaan Hirsi Ali and Theo van Gogh in 2004).

We choose our government and the policies that we want to follow in the Netherlands. Not that you as an individual can determine what will happen, but we all agree that this is a good solution, which makes us feel comfortable. Of course it would be nice if some referendums would be held and we would get the chance to let them hear our voice more often. But that also has its disadvantages, because the organization of a referendum will cost money, lots of money. A referendum can also lead to the situation where the confidence in the government is under pressure. The stability of a country does not usually benefit from such a situation.

You can go where ever you want. You can say what you want, as long as you are not rude or hurting people on purpose. This idea is very good for most people. That gives me the opportunity to write and publish this book without having to fear for my freedom or life.

At home in the Netherlands
The Netherlands is a nice place to live, especially when you are living in the neighborhood of friends and family. I was born here and feel at home. I like to travel to other countries, but it is always nice to come back home. There is nothing like my own house and my own bed. And a hug from my wife and children make my day.

4.4 Conclusion

In the preceding paragraphs, I outlined a number of developments, which is indeed a positive contribution to our society, namely the Environmental Friendly Business/People Plant Profit and the private quest of many to the spiritual in life. We are becoming more aware of ourselves, the world around us and the impact that we have on the Earth. The first focuses on the prosperity aspect of our society and the second on the welfare aspect. If both developments may acquire a strong position in our society, we can probably make the transition to the spiritual society.

Despite these positive developments, the good reader will have noticed that there a critical undertone still exists. We are well aware of the problems in the world, but the capitalist system we use, does not admit definitive solutions. The system is based on more and more prosperity, more and more production and in addition needs more and more people for production and consumptions purposes. This principle is not balanced and certainly does not fit the current developments of the environment among others.

It might sound crazy, but the ultimate solution is growth, not primarily in terms of wealth, but in terms of humanity. We as people need to develop in a positive way. In this way we learn to understand how things can be and should be done differently. We will make the transition to another social system, a new society, where we as human beings and as a society are able to grow once again.

In most western countries, such as the Netherlands, the majority of people have a good life. That does not mean that there is nothing to improve, because ultimately, we want everybody to be happy. A majority being happy sounds a little abstract and does not address very unfortunate individuals, even though they are a minority.

To prevent ourselves from going further down the hill, we must continue working on the positive developments. We should keep searching in that direction. When we do so, we will see a number of concepts coming back:

1. balance
2. environmental awareness
3. welfare
4. spirituality

And of course we need a certain level of prosperity to give meaning to these concepts. When we work towards this and combine it with the conclusions of the previous chapters, we will automatically go into something like the spiritual society.

Chapter 5

The concept of the spiritual society

5.1 Introduction

Good and bad, love and hate

Now we can start with fun part of this book, under the motto: 'Things can only get better!' Because that is what I believe. All human beings have a good and a bad side. After a review of all the bad sides, I would like to appeal to the good side in this chapter. I do not think that people are consciously choosing for the bad options. I think that everyone, who is psychologically sound, wants to lead an honest and happy life in harmony with their fellow men. That is possible if we want to give each other that opportunity, with the emphasis on "want to give each other", because it is really possible, we only have to grant this to each other.

It is also a question of giving love in your heart a chance and excluding the feelings of hatred and jealousy. Love and hate are opposites, which of the two will prevail, is strongly influenced by how we interact with each other. We also have the choice to awaken one of the two in our fellow human beings. When we all have love in our hearts and try to stimulate this amongst our fellow men, you will see that evidently a more loving society will arise. It is something that will almost happen by itself. Can it be more beautiful?

I must also admit that if the bad side of mankind eventually wins, this would be a big disappointment to me. I think we all want to educate our children well and give them positive ideas for their lives. Let us make this possible. Encourage the good and give the bad no chance.

Working together for the future

We are all part of a larger whole. Everyone's contribution to the whole is small but not insignificant. The effects of our actions cannot always be predicted in advance. I hope that I am doing the right thing by telling my story. I want you to know that I only meant well. I have the feeling that this is a message I should bring as my contribution to the world. And regardless of what happens next or how it is evaluated, I feel happy and complete as a human being with a positive state of mind. This makes me realize that things can and should go differently, there is a way out, there is a future. I learn a lot from all of this. I also make mistakes and have my weaknesses. I must also learn to advance on my long journey through the universe, wherever it may take me. The energy that is inside each of us is never lost, but do you want to light up like an angel or do you prefer to be an ally of evil? If you choose for the latter, you have a very long and hard road to go along before you will see that the eternal love is always stronger.

The spiritual society as an example

This chapter provides an image of a society that does justice to the existence of every human being on Earth and is aimed at interacting with each individual in a respectful and loving way. Be fully aware that this is meant as a possible example and not as it should be by definition. Other variants are also possible. This example is only intended to help you imagine what it could look like, one example that can be achieved, but only if we all choose to do so. Nowadays we often blame the system for all the unfairness, we say that this is due to evolution and the way the power is divided in the world, as if we do not have any influence. Let me say this about it:

A. We do have a choice, and
B. Acceptance of an unfair system and participating in it, what does that say about yourself?

With the starting point being love for each other and to cooperate are better than competing, a new society looks very different from the current one. If you also take the money away, you will understand that a lot of things are different in this new society.

In this chapter, I try to describe the spiritual society as an independent society, without taking into account countries that still maintain their current system, but with which there is still a relationship and/or dependency. Chapter 6 describes a migration path, which takes such countries into account. I think this is the most real situation. There will always be countries that will only follow once they have seen that it really works. If we elaborate on these countries in this chapter, the essence of what I would like to make clear will be clouded over, and the process will become unnecessarily complex.

Essence

I will examine various sectors of society to make the necessary changes clear. I do not want to suggest that this is a complete picture and that I have thought of all possible aspects. I expect to get additions over the course of time. The most important is the essence of the idea. Everyone's life is worth the same and therefore everyone has an equal right to a worthy human life. This prevents many problems, as we know it. I am not saying "an equal life", because each person is unique and has different individual needs. People are in different stages of life and every person develops differently. The key lies in the fact that everyone should be allowed to make choices that reflect their personality, skills/capabilities and dignity as human beings. This is not only better for each individual; this could bring the society as a whole up to a higher level in the long term. It should enable us to deal with problems better, which every man and every society is facing. Problems will no longer be permanent in themselves and solutions will not only fight the symptoms. Life will be much more fun for everyone and there will be so much more to learn. Perhaps life will even start to look a bit like heaven, maybe beyond our wildest dreams.

Thinking outside existing frameworks

Many readers will have to get used to this description, to put it mildly, because it is difficult to think outside the framework of the existing system. It is a completely different way of thinking than most us are used to. Many people are also still convinced that the current system will come with solutions or have already abandoned hope and do not believe in the strength of the positive. The people I am talking about are unjustified optimists and real pessimists. Do not think that I blame them for thinking that way. I only hope that this book will get them thinking and perhaps change their minds.

This new system can only be understood if you are willing to let go of what you know and what you possess, and it's important to remember not to reject the new system at the first sign of weakness. You must be prepared to think positively and to work towards a more future-oriented and sustainable society. Of course it is easy to reject something if you want to, but remember that time is running out for the people on Earth. If we continue with the current society, we can expect measures in the coming years, which will impose significant limitations on our lifestyle. Moreover, it is debatable whether they will really help. In the spiritual society, the opposite is the case. It will increase welfare for everyone and everyone can and will lead a richer life than ever before and all of this in balance with all life on Earth. Isn't that what we all want? Are you getting curious about what this spiritual society will look like? I shall keep you no longer with my introduction. Read the following paragraphs and let the essence penetrate you. Try to imagine what the spiritual society can mean for you and your family.

The next section describes experiences of someone in a spiritual society. This is an example to show you what a spiritual society would be like. In the paragraphs thereafter a more theoretical treatment is written including the various components of the spiritual society.

5.2 One day in the spiritual society

I woke up this morning and got ready for work. Whilst brushing my teeth, I stood there and it occurred to me that not a lot had actually changed since the big change. The biggest difference is that the world of today has become a great place to live for everyone. The people are really nice and willing to help each other without asking for anything in return. People do not immediately think about money anymore if anything should be done, but more about the reason why something should be done and the impact it will have on other people and the world. There is no war and crime anymore. People have more respect for each other. We are more aware and care more for each other. It is almost invisible, but it makes a world of a difference.

Respect for each other

People are greeting one and other, because they have respect for each other and want to show they are pleased with others being there. People are more likely to have a chat with each other, and take each other into account. Horses do not drop their waste on the street anymore and cats no longer in other people's gardens. Dogs go out on special designated areas, and children can play on the grass again. People give each other priority in traffic, open the door for one another or give their seats to the elderly while on public transport.

Provide one's own energy

We have lived in the same house for many years. A bigger house is probably possible, but that would also consume more energy. Moreover, we have enough space in and around our current house, even though it is a house at the end of a block in a city. The house is equipped with wind mills and solar collectors, with which we generate our own energy. In addition the house is far more insulated and thus takes less heating. Our car runs on the same kind of energy, so in retrospect our house is a gas station. Therefore, we never use energy from the main suppliers, but we produce all our own energy.

To 'buy' / build a house
It is a lot easier nowadays 'to buy' your own house. You can see at the town hall which houses are empty or are being released within the next six months. If there is nothing you like, you can also choose a piece of land and build a new home. When a house is empty for over a year, they are demolished or rebuilt until they meet the requirements of future residents. The houses are therefore less boring, because everybody gets the chance to express their individuality. More freedom has been given to people's creativity, and they have more opportunities to express themselves. Furthermore people do things in a more civilized way, because they have better education and training than previously.

Car ownership and use
My wife also works, but we only have one car, because we work and live in the same city. The distances are short. A second car would also put a strain on the environment. We will keep this car for as long we can, until it is irreparable or doesn't fit in with our family situation. The production of a new car demands a lot of raw materials and there will be a considerable amount of product waste. Furthermore it is not responsible to renew your car every few years.

Companies and products
Yes, there is a substantial difference with the capitalist system. We now produce on a need basis and you do not need to pay to obtain products. The company has therefore become a society in itself. People have pleasure in their work, the atmosphere is very good, the workload is much lower per person than it was in the past, people take time for a pleasant chat and help each other if something goes wrong. Sometimes there is just a little stress if something has to be delivered on time, but people see that as a challenge. The planning is made together with the people. The production level is the same as in the past, but that is because people find it fun to work here, which causes lower absenteeism. Employees feel more at home nowadays and feel involved in the

company. They also like to think more about things that are important for the company, which helped us to take different efficiencies and environmentally friendly measures. We now produce more efficiently and cleaner than ever before. Of course there is always someone who, after a period of time concludes that he or she wants something else that we cannot offer. Then we call someone from a special institute for retraining and place those people in another company. This does not affect a person's personal situation, because you earn the same in any job whether you are a street cleaner or a director. The same institute also takes care of us by finding a suitable replacement. So we do not need to advertise and do not lose much time and energy on recruitment.

The point is that you have the chance do whatever you like, within your abilities and that helps the people around you. Therefore everyone works with great pleasure and commitment. That is what you are rewarded for. All your working hours are being administered through the internet in the central database containing all the people, which all companies and institutions are connected to.

Production and raw materials

Our company provides their own energy for a large part, because we have large wind turbines and the entire roof of our building is covered with solar panels. If we have power left over, we give it back to the electricity company and if we need more, then we just tap it off.

We need raw materials for our production. We order them from a supplier, when they come from a country with a spiritual society, there is no problem. Thus we have long-term agreements for the supply of materials. When it is a supplier from a country with a different kind of social system, then the supply goes through the government. Fortunately, this is increasingly less the case, because there are more and more spiritual societies. They are providing each other with more and more essential materials.

Means of payment

When I go to the store for shopping, I take my ID card along. This is an identity card with which you can identify yourself in combination with fingerprints and an iris scan. Contact is being made with the central people database, in which all personal data is firmly established. This database is in turn linked to the system that keeps track of how many hours you have worked for the society. This is the new base for the payment system. When I need 'to pay' for something, there is a connection made with the central computer and it checks whether I have contributed enough hours last month. I can take all the products with me if I have worked enough hours. The working hours are not taken off, the registered number of hours stays the same. In the next store this will be checked again and I can take everything away with me again. I will not be greedy in what I take, because it makes little sense to take more than you can eat, for example. Moreover, all products go rotten quickly, because there are far fewer preservatives in them than before. All vegetables are produced in a biologically sound manner. And finally, it is antisocial to take more than you can eat and to throw the rest away.

Restrictions

Limits have been set to a number of products; this is to keep tabs on certain products and people taking things that are not strictly necessary. In principle, for example, every family has one car. If you want a second car, you will have to put your argument across as to why you think you need a second car. When you have good reasons, then you will generally get that second car. But sometimes a different solution is suggested, like a scooter, a motorbike or moving to another town in the vicinity of your work. A second house is never given to someone. You can only live in one house at a time. To leave a house empty for a large part of the year is a waste of land and raw materials. There are vacation resorts however, but they are occupied all year round, as the facilities are all-season. Think of sports halls, indoor playgrounds, shopping malls, indoor tropical swimming pools, indoor ski slopes and indoor Go-Kart with electric karts.

Sometimes products cannot be delivered immediately, due to high demand. Then you sometimes have to wait a few weeks or months until it can be delivered, but that has always been the case.

Holidays

I book a vacation, when I want to go somewhere on a holiday, I have to work a minimum number of hours per year. In addition, a number of weeks remain free, in which I can take a vacation. We have no more central school vacations, so you are much freer to book your holidays. A family could do less with their money in the past, than people without children, because you also had to pay for your children. Plus you had to take a vacation in the most expensive weeks of the year. Now it costs nothing and the school is open all year round. The parents can plan a vacation whenever they want, without having to worry about the busiest periods. There are no massive population movements anymore or fun parks that are overcrowded. Life has certainly become more relaxed.

No crime

A big advantage of a life without money is that there is no crime. It is of no use to steal something. If you want to sell what you have stolen, you would not get any money for it, and if you did you couldn't use it. Furthermore people can get things free of charge; nobody wants to buy something from a thief. To sell drugs is also of no use, because people are giving no money for it, far less people will want drugs, and the drug trade will disappear.

The value of an item is therefore not relevant, but it is about what you do for society. All material things are really worthless; they are tools that make life more pleasant and easier, but are not the 'be all and end all'. That is only the breeding ground for criminal activities.

There is also another reason why there are no criminals anymore: they are all working. What makes a man become a criminal? This mainly occurs when someone is not able to build a wealthy existence with his or her capacities and capabilities. In the spiritual society, anyone can have a prosperous life, regardless of his or her capacities and/or capabilities. That is a fair basis and provides an equivalent existence for everyone. So why would you still want to be criminal? Doesn't that sound good? Never again having to be afraid that your bike or car might be stolen, your house broken into or that your shop is being robbed, and no more bags robbed from old ladies.

Care of the elderly

The care of the elderly is fantastic nowadays. There is much more respect for the elderly than in the past, because the government mainly consists of older experienced professionals. They are held in much higher regard, making young people to be more inclined to contact the elderly and to help them if necessary. There are enough nursing homes for the elderly and opportunities to interact with other people to keep people busy. These centers are set up by the government. The staff has much more time for the elderly and work with a brighter attitude. The people in the centers also notice this and it makes for a brighter atmosphere. Moreover, there are still a lot of elderly people living in their own house, this is possible because home care has improved a lot. They have received more resources and more time per employee to spend for the needy people.

Education at school

My children like to help their grandpa and grandma. They were taught this at school. Values and standards are an important subject at school. The schools are less focused on performance and more on the child. The things that are right for the development of the child, is always important for society. People, who feel at home at school, will also feel more at home later in society. The school tries to prepare children for society and to find out what a child wants to be, moreover what his or her talents and interests are. The study of the universe is an important new subject in school. During this subject the children are taught how the universe works and what the impact is on daily life. It is presented as a scientific subject.

Religion and religious communities

Another thing that has changed is that we often visit a church or mosque on a Sunday. Science has researched the existing religious and spiritual authorities for a common basis for the benefit of society. They have come to recognize a 'divine' greatness or entity, where we as people, are part of or subordinate to in a certain way, and that love is what connects us all. This is now propagated by all religions. There is more knowledge about the universe, and understanding of what this knowledge may mean for our society. The existing religions have become more aware of the fact that through cooperation they can play a significant role in society and can therefore make a greater contribution. Of course, there are still differences between religions. This will never change, just as there will always be differences between people. But there is more understanding of each other's positions and because of more meetings and cooperation, the differences have become smaller.

We have become more aware of the 'divine' in the universe. We sometimes have a session with a psychic, who not only shows something of his talents, but he also explains how he sees and experiences this. Science is also paying more attention to the paranormal phenomena because it is seen as a part of the key to the universe that we still do not fully understand.

The religious and church communities in our district are jointly executing a wide range of tasks within the neighborhood. They are running a combined children's club, where children can do creative activities after school. They are taking care of an additional piece of home care for the elderly in particular. These are not the standard care activities, which are executed by the home care centers, but activities to keep the elderly involved in the community, and to give them a feeling of involvement.

In addition, staff are regularly giving presentations at schools and events to elaborate on the universe, how it is functioning, our place within it, and to disseminate this knowledge and keep it alive.

We ourselves are helping with a number of activities. Actually, the church has become more of a community center, where residents can meet and talk about issues that go beyond the daily concerns. There are also a lot more people coming there than there were before, because the church has loosened up a little. The essence of religion is what it is all about today and what we can do with it in our daily life. Everyone is welcome and it is always very pleasant to go there.

Helping developing countries

Finally, I would like to say something about helping the developing countries. Previously, we only offered help, because we felt, as "rich" Western countries, that we were required to do so. In this way, we freed ourselves of our guilty feelings. They were helped as far as money allowed. Now we help these countries, because they are people and children, just like you and me. Now money is not around, so it's not about if there is enough money to help but it's a case of if there are free hands available then people will help. The aim is simply to help everyone in need. We do not want every country to be the same as an average Western country, but to help the developing countries to grow in a way they feel comfortable with, by advising them, consulting them in any way we can. What steps are taken and how big those steps are may vary from country to country, as well as the final result. It is primarily about creative ideas, exploring possibilities with no limitations. In addition it is about positive thinking, striving for a healthy lifestyle for everyone, a life in which we can all be happy. We want people to develop, we want to give them a form of existence in which they can live healthy and happy in balance with nature.

In the spiritual society, we can really do something for these people, money is no longer an excuse for not helping.

Available work force

There are more hands available, than there were in the past during the heyday of capitalism, because a lot of work with a financial character is no longer needed. Companies still have an administration, but only a stock administration, this is much easier than having an additional financial administration. Many financial institutions are no longer needed, such as banks, insurance companies, audit firms, tax advisory offices, and investment companies'. Many hundreds of thousands of people previously used to work with these companies. All these hands are released to assist in the care sector, construction work, developing countries and with companies where there is an increased need. The garden centers are a simple example. Many more people can now have a beautiful front and back garden for their house. But there is also a growing need for old people's homes and restaurants. They will get a lot more customers. In short, a huge shift in society has occurred. Life has become much more pleasant and sociable.

Closing of the day
I usually close the day with a meditation session of about fifteen to thirty minutes. During this session my body relaxes, my mind comes to rest and I get into contact with the universe. As a result I experience a deep inner happiness. After that I am able to sleep well.

I hope that the above description has shown what a spiritual society could look like. In the following paragraphs the various aspects of the spiritual society are described.

5.3 Religion and spirituality

Do you believe in a 'God'[1]? Do you believe in life after death? More than 80 % of the world population believes in a 'God', a universal essence, a 'divine' or universal entity and/or a life after death/reincarnation. This means that a large majority of the people believes that there is more between heaven and earth, than purely the physical earthly life. They recognize that there are forces, which are larger than ourselves, which are giving direction to what is going to happen to us after we have died. They generally believe that love for each other is the most important thing there is on Earth and that we must help each other. Isn't it time that we start to live as we believe? That we will help each other?

---

1 The word 'God' must be understood in the broad sense of the word. There is no referring to one specific god. Each described non-physically Supreme Being, entity, energy or force that can be considered as the most essential connection in the universe is covered by this word.

Universal love

A universal essence of love that wants the best for all the people on Earth, it would certainly not want them to fight over his word and will. Love is a feeling, not something of the common sense. The more you think about it, the farther away you get. Search in your heart for the positive and you will find love. Then you will see that that feeling is not only pointing inside of you, but also to the outside, to your fellow man. Keep hold of that feeling and think of what you should do. Soon you will come to the conclusion of helping others. That is where our common sense starts, because what is the best way to help others? How can you mean something to someone else? That is what you will need your common sense for.

Is this difficult to understand? Who would not be able to understand this? It is not my intention to rewrite the word of a "God" or talk like a theologian. This is purely an attempt to search for the essence of every religion, the core of the universe. With all my feelings and intellect this is where I end up. This is what I would like to tell you, and from the depths of my heart I really believe that I am doing the right thing. The solution to all problems is in my opinion, in positive thoughts and feelings. The solution is not in violence, in war, in the eradication of people or the banning of anything, but in the making of choices and raising awareness along with educating people in these choices, so they will start to understand and to live up to it. The solution is also in simplicity: love is simple, for everyone to understand. Everyone wants to be happy and not get lost in a jungle of regulations. Rules and laws have become an end in themselves, while they are only tools. We should focus on each other, not on money, power, status, rules or laws. It is of no use to discuss endless details. It is wiser to use your time and energy to cooperate. If we do that, the society will become much simpler and there will be a lot more chances to make things better for each other, here on Earth.

Belief in a universal entity

Back to the belief in a 'God' or a universal entity. If almost everyone still believes that something like this exists, it seems important enough to me to make that a central theme. In order not to do so would be hypocritical: that sounds like serving two gods, in my opinion that is no good. On the other hand, it makes things clear: we believe in a 'God' and are living up to it. In addition, it also solves the power problem: there is only one real power: the power of the universal entity. Everyone and everything is subordinate to it.

The building of power, control over others has become incorrect. An important assumption is that we have learned from the past, in which knowledge of the religion is often misused to oppress others. Everything we do here should be focused on helping others, even if that other person has done something wrong. Perhaps this does not sound much different than it is, but if you have a different outlook, more positive, then you know that it will be different.

Looking in a 'different' world

Near-death experiences and paranormal gifted people regularly offer us a glimpse beyond what we can observe with our primary senses (eyes, ears, mouth, nose and hands). This helps us to look across borders and allows us to see that there is still something more to understand, that is more powerful than anything we know in this earthly life. There are people who still have doubts. But how much proof do you need, I wonder? We do not have to wait until everyone is psychic gifted or has had a near-death experience, I hope? The most important thing is what we can do with this information. Speak with people who have had such an extraordinary experience. Overall these are positive experiences that have given people access to universal wisdom and have often led to a turn in the lifestyle of these people. Isn't that more important than actual scientific evidence?

Learning from experience

Again I would like to emphasize that it is not my intention to recommend the consulting of the spirits. There are also ghosts with a negative attitude, wandering spirits who have not yet found the path of universal love. Contact with these spirits is not without risk. To be a psychic is a gift. If you are one, it is almost impossible to ignore it. Use it in a positive way. But do not start deliberately looking for it. Everything in life has a purpose and when the time comes, you are automatically brought into contact with it or not.

As humanity we can learn something here, this is a confirmation of a little piece of the universe that can help us to understand how everything works, why we are here on Earth and which can give us direction in our earthly life. We should not hide or ignore the existence of the things that we encounter. By doing that we limit our horizons, which could result in taking wrong decisions, because certain facts have been left out which prove to be essential later on.

Learning through inner experience is the most important way of learning. How do I get into contact with myself and everything I stand for as a human being, but also as a universal being? In this way we will be able to learn more about ourselves, but also to understand that we are all connected and that love is the binding factor. Meditation is an important tool. I will refer to this in the section about education in school.

Giving meaning to life

A third major concern in that this context is giving meaning to life. Everyone comes to a point in life where you ask yourself what is the meaning of life? What on Earth am I here for? Why am I doing what I do? Atheists will soon come with the answer: just to enjoy myself as much as possible, because after this it is over. Many people with a religion often have trouble finding a good answer, because they are waiting for a sign from above.

Giving meaning to life is probably the most important thing in our lives, because it focuses on the essence of the existence of every individual. Isn't it time that we start doing something fundamental to give meaning to our lives? Why don't we help our children with this? Why do we make everything more complex than necessary? Why are we still unable to explain how life works? Do we really listen to each other? Do we listen to elderly people with a lot of life experience? Do we really listen to our loved ones? Do we listen to the universe?

In short, we need this knowledge to be able to focus and to be really aware of life, a kind of navigation system, to make us realize what we are doing. The road is long and there are many side roads. It is a pity when you are running out of fuel, while you are actually just trying the side roads. The highway and the warm reception at the place of destination are achievable for everyone. Let us help each other to get on this highway, and to reach our destination.

Giving meaning to life takes place, because knowledge about the universe makes people understand what is important in this life: love and happiness within yourself and for your fellow man, but also an understanding of why we are here: because death is not where it ends. This understanding leads to a form of consciousness, which also can be acquired in other ways, such as meditation or a near-death experience. In that respect, there are several ways that give the same result.

The problems you are dealing with here on Earth are still there after you have died. You will carry them with you, until you know how to deal with them in the right way. This is not intended as a punishment, but as part of the learning process for you as a (part of the) universal essence, to bring you to a higher level of existence, to a level from which you are able to help others in the universe. Where that eventually should lead to, I do not know. I am aware that I do not have all the answers. Most importantly by giving meaning to life people get a sense of direction and are contributing to society with more pleasure than ever before and want to make it a fine place for everyone.

Spirituality

In the preceding paragraphs, I have named three important themes, all three have a close relationship with each other and the lives of people on Earth: spirituality. These themes are: 1. belief in a universal entity and/or life after death, 2. learning by experiences from outside and inside yourself, 3. giving meaning to life. These three can also be read as love, knowledge and awareness. In the lives of people we can, in addition to a physical and mental component, also discover a spiritual component. These are, in my opinion the three components that form the basis of human life on Earth. It is time to start working seriously with this last component in society.

At the beginning of human existence the focus was mainly on the physical welfare of human beings. Providing the most important resources for life was all they did in those days. This is still the case in many primitive countries. It was only after we came to the division of labor and 'automation' of many simple, especially manufacturing jobs, there was more emphasis on the mental component of man. Psychology, Sociology and Education became important as 'soft' sciences. Understanding of how people are mentally functioning offered opportunities to help people in other ways. It also helped to complete the picture of how people are functioning as a whole. As the concept of prosperity had been sufficiently researched, there became room to pay attention to the welfare of people.

Parallel to this development, man has always believed in a 'God' or something larger than himself. We have given it a place in churches, temples and mosques, but where did we go from there?

The fundamental needs of human beings

Then it became quiet, because materialism became the new 'God' on Earth. The golden calf has been worshiped ever since. However, the psychological need of humans to develop themselves spiritually cannot be denied. There is support for it in the scientific world. This is nicely symbolized by the pyramid of Maslow (scheme 2). Maslow was an American clinical psychologist and founder of humanistic psychology.

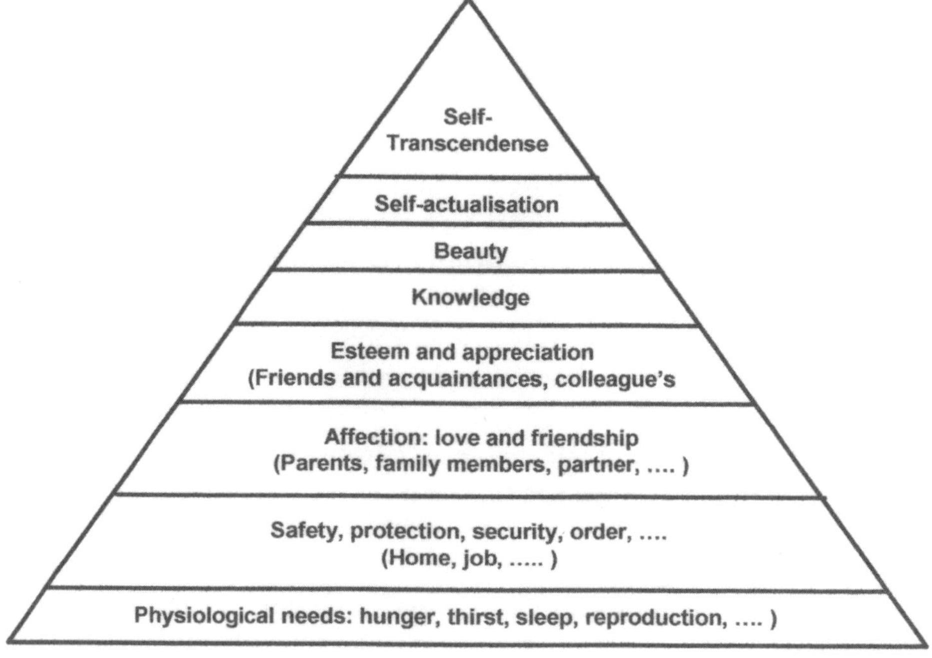

Scheme 2. The pyramid of Maslow

According to Maslow some basic minimum needs must be satisfied to be able to develop a healthy human personality. These needs are congenital and he classifies these needs as follows:

1 Physiological needs: these are the basic needs such as eating, drinking and sleeping, to make sure that there is a physical balance;
2 The need for safety and security concerning job, housing, relationships, etc.;
3 The need for social contact, friendship, love and positive social relationships. In Western society increasingly less can be contributed to these needs. Typical examples are: the increasing ego centered way of living and the increasing loneliness of modern humans;
4 The need for esteem and appreciation, which increases the competence and prestige in the group; to attach importance to the status in social context;

5 The need to develop his or her own intelligence and the acquisition of knowledge to be able to understand the world around oneself, as part of self-realization;
6 Learning is stimulated by looking at the surrounding world and learning to appreciate the beauty of it. Innovative insights and to behold beauty stimulates the aspect of self-realization;
7 The need for self-deployment or self-actualization is the need to develop one's personality and mental growing possibilities. Trying to get the maximum out of it. The social environment is essential as a support base of this actualization trend;
8 Finally, Maslow recognized an even higher level: transcendent experience, also known as spiritual needs. It is about building a relationship with the non-physical world and the understanding of questions, such as the purpose of life. It also has to do with the attitude towards life as a whole and all the related goals and cycles.

Maslow described people with a high level of self-realization as follows:

1. High level of consciousness (open for reality);
2. Striving for honesty (a democratic character structure amongst others);
3. Targeting for freedom (spontaneous, creative, being 'yourself');
4. Radiant confidence (accepting yourself, others and nature).

Maslow discovered that healthy individuals are motivated to work on what he said self-actualization and described self-actualized people as people who can be characterized by corresponding characteristics. He described self-actualization as:

"an episode or spurt in which the powers of the person come together in a particularly and intensely enjoyable way, and in which he is more integrated and less split, more open for experience, more idiosyncratic, more perfectly expressive or spontaneous, or fully functioning, more creative, more humorous, more ego-transcending, more independent of his lower needs, etc. In these episodes he becomes himself, more perfectly actualizing his potentialities, closer to the core of his being, more fully human. Not only are these his happiest and most thrilling moments, but they are also moments of greatest maturity, individuation, fulfilment - in a word, his healthiest moments.

Self-actualizing people, those who have come to a high level of maturation, health and self-fulfilment, have so much to teach us that sometimes they seem almost like a different breed of human beings."

During later research Maslow came to the conclusion that every man is clamped between the basic needs of conservation and the growing need for new experiences. To be able to grow, one must reduce the attractiveness of the 'safe situation' and the danger of 'growth'. With this 'safe situation' Maslow meant the security of the original situation. Man is now established with his habits. In order to change these habits and to get new experiences, one has to break through this 'safe situation'.

Empirical studies have found little or no support for the ideas of Maslow. That does not mean that the identified needs do not exist, but merely that these cannot be organized in the way Maslow did[2].

In search of vision

We are getting stuck in many areas in our current society, because we do not understand what on Earth we are doing. There is no policy that fits the operation of the universe or life on Earth.

In the spiritual society, we accept that everything hangs together. All your actions have consequences, which are the responsibility of yourself. You can not refer to ignorance in a world where all the knowledge is at your disposal. You cannot ignore the suffering in the world, because it is there and you can choose to do something about it. It is your choice!

More and more people in today's society go in search of a long-term vision and arrive at the aspect of spirituality, finding tranquility in oneself, searching for balance between intellect and emotion, going past the boundary between consciousness and sub consciousness and getting in contact with the universe. It is an underexposed aspect and they are right, the solution to our problems lies here, that is why the next step should be the spiritual development of mankind.

---

2 The above text about Maslow is taken from the Dutch and English Wikipedia sites on the Internet.

Cooperation with science

Religions have been around for a long time, actually, as long as there have been people on Earth, but this has never resulted in collaboration with science. Contrary to religion, science is being seen as one of the fundamental pillars of our society. Science has always thought of religions as being unfounded. There has always been a situation of opposing sides: can we prove that the flooding of the Earth never happened? Can we find a body of Jesus Christ? Can we prove that 'God' did not create the world in 7 days? But is this really important? We are talking about things that happened long ago in the past. You must learn from the past, but it is not necessary to know all the details. We now know that there has been a flooding of the Earth (see The Epic of Gilgamesh), there is growing evidence that confirms the Bible stories and we get more and more knowledge of the evolution of the Earth. But are the theories of evolution and the Bible incompatible? The Bible was written by people, just like this book. I am trying to write this book, so that everyone can understand it. Some things described or designed will be too simple. But it does not make the message less true. You cannot fill in the future by exploring the history or even deny it. History can teach us why we are who we are today, but for the future, we must look to the present and take into account all the knowledge we have, to take decisions. These decisions do not have to be extrapolations from the past. We have got brains to choose a different path, or to leave out things that we previously did, or to think of something completely new. However, it is time for us to implement whatever keeps most people busy on Earth in a positive way.

Science serving humanity

Coming back to the aspect of spirituality and science. Science is not an isolated thing, but it is there at the service of humanity. Therefore science can not show us the way we should go, we must do that ourselves. Science can only support us. What we need now is a complete understanding of the universe, so we can understand why we are here and in which direction we should go. We naturally come to the aspect of spirituality that can connect all the people on Earth with each other. Science can build a broad foundation. The foundation that we need is a description of how the universe functions, so that we do not have to fight about this anymore and everyone knows where they stand. I think this is the next major challenge for the scientific world. Science has to go further than ever before and perhaps even use reverse reasoning, because a link has to be built between the physical and non-physical reality. Maybe science should state that something is true, based on the current information, until the contrary is proved. I want to leave the honor to science itself to determine which rules that information should comply to.

Description of the universe

Science gets the assignment to provide substantiation. This means that there should be a 'scientific' research for paranormal phenomena, near-death experiences and religions. Scientific evidence through observation is not what we should be looking for in the first place, but rather to establish relationships and parallels. With this we should be able to assume that we are dealing with the relevant issues with a sufficient degree of certainty, which can justifiably be regarded as the basis for our society and religion. As a result this should lead to a model of the functioning of the universe. The most important thing is that we, as humanity, can make another step in our development process. We must not stand still. This is only possible if we look beyond the past. We are going to slow. If we want to save humanity and the Earth, we will have to move faster. Clinging to what we have will not help us. There is enough knowledge, but we will have to connect this in order to create the complete picture of the universe, this will ensure wide support. It is only in this way, that we have a chance of building a basis for the spiritual society.

Example of a description of the universe

Based on all the information that I have gained in recent years, I would like clarify the previous paragraphs a little by giving an example of what a description of the universe could look like:

Within the universe, there is a fully connecting and loving entity, an entity that stands above life and death. A greatness that is always there, always was, and always will be. An entity of infinite love, where there is place for everyone. We, as living humans in this universe, are part of or subject to this greatness. We should go on developing ourselves just as long, until we finally are allowed to take place in the world of this greatness. When we are there we have no shape and no senses, but exist as a unit of energy, as a universal entity. Therefore, we are all equal. Some will call this heaven, others the mecc, etc. To be up there, we need to follow the path of love. The Earth is the place where we can learn. When we die, we become what we are in essence, a universal entity. We are asked what we have learned, and if we think that we have given enough love to others. If not, we return in another existence. The existence we had, or footprint of our past life, will not be lost. Everyone is therefore in a different stage of development. We will come back again and again, until we have learned enough. This is what makes the Earth for some a place almost like heaven, while for others it looks more like hell.

We come from the universal entity and will go back there again. On Earth, we have to find our own path. We will come into a place where we can learn more given our stage of development. We have been given our mind and emotions to help us to make choices. We learn from the choices that we make. We can only blame ourselves and our fellow man for what happens to us as people. It is intended that we learn to make the right choices. We need to help each other. That we stop and prevent diseases. That we avoid and prevent (natural) disasters. That we fight and prevent poverty. We need to take responsibility and take action.

We cannot learn without making mistakes. If everything was all right, it would have made no sense to put people on Earth. This does not mean that the universal entity approves of evil. It is not intended for children to die young, or that we suffer from pain. Eventually he wants everyone to develop in a positive way to be able to come to him and to exist in love. The universal entity does not condemn, but gives you insights, so that you can learn from them.

The above is just an example and not intended as the ultimate truth, it is a description that everyone can understand and can be adequately substantiated in the long-term. This is the first challenge we face: a description of the universe that can serve as a basis for the spiritual society.

The purpose is to make it clear that we need some understanding of the functioning of the universe to integrate into our daily reality and life here on Earth. With this knowledge we can be more effective in what we do and use our energy of life in a more focused way.

Principles of society

It is important that there are common principles in a society where the vast majority can give support and that a small minority has no problems with. An important principle in the spiritual society is the belief in a universal entity, whatever that may be. In the end we are all connected to this entity and to each other. The binding factor is love. In the spiritual society, we no longer debate about the existence of a 'God', but we explain the presence of a universal entity to the primary principle of the society. The same goes for life after death. The most important aspect of these two points is not whether they are or are not proven, but what conclusions can be drawn for the way of life in society. In the first place this will make all the current positions of power on Earth subordinate to something higher and in the second place create room for something like giving meaning to life. Power is no longer meant to be used for self-glorification, but subordinate to the universal entity. Actually, we no longer talk about power, but about governance. The executive board on Earth is there to help the people and is not standing above the people, but under the people. The executive board is only there to organize and coordinate things, because otherwise a society will not function. All the actions from the executive board should be made from this perspective.

The executive board on Earth should apply a number of basic rules which we all are familiar with, as they are equal for almost all the main religions and are of a nature that they will almost certainly be acceptable to every individual. To makes things clear here are some examples these basic rules will follow:

1. The Earth is our mother, care for her;
2. Take from the Earth what is needed and no more;
3. All life is sacred; treat all living beings with respect;
4. Honor your relationships;
5. Speak the truth, only of the good in others;
6. Love your neighbor as yourself;
7. Do what needs to be done for the welfare of all;

Specific from the point of view of the spiritual society there could be added:

8. Show ethical behavior (do not cause envy, ensure a balanced distribution of wealth, educate yourself and contribute to sharing knowledge);
9. Everyone should contribute to society, taking into account one's talents and possibilities.

This is another example which indicates the direction that the spiritual society will probably develop. It is not my intention to say that these should be the basic rules. Perhaps I missed something or I have formulated something completely wrong. It is only intended to help you imagine what a spiritual society could look like, and to contribute to imagining a better society.

Scientific study of the universe
Aspects that lead to too much discussion should be left out of the picture. A permanent institution should be set up to continuously do research into the functioning of the universe beyond the boundaries of life and death. They can bring a better understanding about the essence of our life on Earth. Without this understanding, we do not know what we are doing, and many people throughout their lives are left with the question, for what reason are they here on Earth.

We are also going to look for the similarities in religion and spirituality among people.

Freedom
Filling in further details of the belief in the universal entity and in the spiritual society is not desirable or appropriate. This could lead to a limitation on the freedom and creativity of people. That is not the intention. We want more freedom! However, we must realize that the path of love leads into a certain direction. To hurt someone on purpose, to murder or endanger someone's life is not love. The rules and laws in society will be adjusted to fit this new situation. On the other hand, we must be aware of the risk of over-regulation, because a narrow road is no option. The more people that are going in the right direction, the less regulations are needed.

Personal development
This does not mean that everyone should follow the same path. That is not possible anyway; based on the uniqueness of each human being and the different stages of development we are in, one man will want to go further than the other. This does not only have to do with the level of development, but also with their ambitions. One man feels he is more called upon by the universal entity at a given moment, than the other. As long as things stay reasonable and logical and are not in contradiction with the universal rules.

The essence of what we believe in the spiritual society is actually something very simple: give love, do well and keep in contact with yourself and the universal entity. You do not have to study for years to understand this. This is easy for everyone to understand. Then you are able to focus on the world around you, on the people and animals that need help. That is where we need to use our energy. When we do this on the basis of this universal thought, the world can be a much better place to live in.

Limit the number of rules and regulations

Universal rules also form a handle. Religion bound rules and guidelines should not be established. Then there is a great risk that they are in contradiction with the rules of the country itself. In addition, one of the principles of the spiritual society is to create as much freedom as possible for the deployment and development of each individual. It is rather intended that the conversation over the universal entity is an open conversation, so that people can learn from each other. To limit the conversation to the essence is good enough for most people, and it also creates a broad commitment. By leaving out all kinds of rules to which people should obey, many more people will experience these institutions as open organizations where they are welcome. People should not be frightened by stories about hell and damnation. Fear is not a good consultant to listen to in life.

The role of existing religions

The existing churches and religious communities may continue their work. However they should no longer be places where the rules of the religion are delved into even further or where people discuss matters like why their religion is a better alternative than the other or where people are afraid of being expelled from the community. No, the existing churches and religious communities have become supporting institutions from which help for the fellow man is organized. Especially small scale help, for example, at the neighborhood level, because the religions are well organized at that level. They will be able to do much more than previously. Also, there will be many more fulltime people deployed (why this can and will happen becomes clearer later in this book). The role of these organizations will also be much more important than in the past.

Of course, the churches, mosques and temples will remain as places where people can meet and where they can talk about the universal entity and listen to choirs singing. Everyone is welcome. The message about the universal entity is to give knowledge and understanding to everyone.

The power of the churches, mosques and religious communities is less over people than ever before. People will feel freer and happier and more willing to cooperate to maintain this new society.

Religious history
Books like the Bible, the Koran, and the Tenach will be used as history books, where we can learn about how religion originated. In addition, this can lead to understanding each other's religion. And finally, each of these books contains wise lessons.

The churches and religious communities will have a supporting role in this area, not in a rejecting way but in an understanding manner. All these books are recognized by millions and sometimes billions of people. Nobody has the right to reject them. It makes no sense to deny them. It is better to try to learn something from them. By talking to people and referring to their common sense and by giving everyone space for their own interpretation, this will eventually lead to a situation in which everyone has found the way in the right direction, the path of love. This creates freedom and openness. A lot more people will feel free to join in and become enthusiastic about it.

Spirituality at any time of the day
In the spiritual society, spirituality is no longer something for after work or on weekends, but something that is part of our daily life. Not in a restrictive form, but in a sense of giving meaning and direction to our lives. However, everyone will have to decide for themselves on which direction they want to go based on this knowledge. Everyone should choose a direction which makes them feel good. There can only be justice done to the individual needs and characteristics of different people, if there is enough space for the individual himself. In this way, everyone can be happy in this society.

Everything in the universe is connected, even here on Earth. Every act should be in line with what we stand for, what we believe in, without any reservation. No doubt everyone has a good excuse for acting differently, but deep in our hearts we know that is not correct. To take action from our hearts will be in everyone's favor in the long run.

The essence of religion is not in what we believe in, but in what we do with it in relation to our environment. The assumption that love is the essence of every religion comes down to loving yourself, and loving and helping one's fellow man. This is not complicated or difficult to understand, this is concrete. Everyone is capable there.

5.4 Detachment

Another important issue within the spiritual society is: attachment. A term which is used a lot nowadays within the spiritual world. But what is meant by attachment?

To others

Attachment starts at the moment that a human being is born. There is a bond between mother and child. It is a form of love. This is a healthy form of attachment.

As a child grows, we start to expect more and more. And if a child lives up to our expectations, we are going to love him or her more and more and get even more attached to this child. When the children have become adults and the parents are still attached to them as if they were children, we can call this an unhealthy attachment. We educate children to be autonomous and independent adults.

It is therefore important that we learn to detach and to build healthy relationships in which we accept people in the way they are and love them, without having expectations, giving love without expecting something in return. By giving people enough room, they are able to develop further.

To things

In the current society we are also attached to money, power, equipment, houses and other things. These are even more dangerous forms of attachment, because by doing this we run the risk of becoming attached to material possessions rather than to people.

In the spiritual society, there is no money. The amount of land per person is limited and in principle the same for everyone. Furthermore, all equipment, cars and houses have lost their value. This means that people will get less attached to them. It is no problem to get something replaced in the spiritual society; therefore it is pointless to get attached to materialistic items.

To one's own identity

Then there is the attachment to one's own identity. Who am I here on Earth? Who do I impersonate? What have I achieved? To which group, class do I belong? In particular this concerns identity related to status. This is the main reason why some people feel more important than other people and sometimes even larger than life itself. This is in contrast to personality, which is about your qualities (of character). How do you interact with and treat your fellow man? The knowledge about the functioning of the universe will make people start to realize that the identity we have is something temporary. It is just something that we need to be able to make progress in our overall development. Once we know this, identity attachment will diminish among large groups of people.

Fewer problems and better solutions

The effect of this decline of attachment will lead to a situation in which we are able to appreciate the right things and the love we are giving to each other will be unconditional.

This will also lead to less mutual problems. People won't feel insulted or like they are being treated incorrectly. Interests and possession have no role of importance to play.

And if there is a problem, there is more solution oriented thinking, and the people involved are taken into account. Solutions will be sought to suit all. No half-solutions, no solutions in the form of compromises, because these are not real solutions.

Finally, detachment will lead to freer thinking to deal with changes faster and easier, these are all developments, which give us, as humanity, the chance to grow.

Prerequisite to really understand what detachment for a man can do is a deeper form of awareness, a sense of universal unification. Meditation may be a very good tool to achieve this.

5.5 Science

Science will have a more diverse nature in the new society. There will be more flexibility and science will take a more proactive attitude. With flexibility, I mean that a distinction should be made to the extent that evidence is necessary to continue as scientific evidence. When we talk about nuclear energy or new drugs, it is important that clear evidence has shown that it works and that it is safe. As long as you do not have proof, in my view you should not, use them. And if you decide to go ahead anyway and find out there are uncontrollable dangers attached, then you will have to decide to stop everything after all, otherwise you are not acting in a responsible manner.

When we talk about Neurolinguistic programming (better known by the acronym NLP), then it is much less important that the relationship between body and mind is proven beyond all doubt. If people are 'thinking' themselves happy by positive thinking and this appears to work for 95% of the people. What would you like to do research for? In this case it would be a sin to put more energy into research than is necessary. If it does not work, it will disappear from the scene anyway.

It is therefore important to distinguish between the extent that something must be proven and the need to do so. In this way we prevent extensive studies of phenomena without a social necessity. There are better ways in which this energy can be used.

A lot of research can also be saved through the establishment of relations between the results of previous research and facts in documents. If it is possible to provide sufficient proof by making these relationships and there is no risk attached to it, it is of no use to do extensive and costly research. Risk management and the balancing of priorities can go hand in hand.

Another important change is the proactive attitude of science. That means taking initiative in starting research for things that are important in the development of society. Science can be a stimulating and guiding force of society, if they can provide information, which the society needs at that time. Science is often done in isolation and occasionally we hear something. We can also turn it around. Society must indicate what they want to know and science must pick up on that, and scientists themselves must also keep an eye on this. In this way, science can play a more important role in society than it is today. For example, popular scientific magazines sell very well, while pure scientific journals have a much smaller circulation. A lot of people are not interested in what is published in these scientific journals.

What I would like to say with the above is that science does not stand alone. Science has the task of bringing humanity forward as a whole. Therefore it is important to know how the universe works, whether we as human beings are able to travel to another planet in another star or Milky Way. But is it also important to know what the average IQ of a chicken is? Or how fast grass grows? We can do research on everything if necessary, but it must have added value. The more energy we put into meaningless activities, the slower we will develop as humanity and the greater the chance that we will not survive in the long term.

5.6 Education

It is clear, however, that this society cannot work on a scientific basis alone. Even the set of rules and laws we have worked out will not produce the desired effect, as long as our knowledge and rules for life are not taught properly to our children in the first place. For adults this only leads to resistance. In fact, if we teach our children what it is all about, rules and laws will hardly be necessary. Living honestly comes from within. The education of children is therefore the perfect time to teach them how to live their life in a way that fits each unique individual.

Education is essential for children to become adults. In the past we have always transferred our knowledge and attitude towards life to new generations of children. This needs time and attention. If we make children aware of what life is all about from a young age, there is a chance that people will start to live according to this knowledge and attitude towards life. This will therefore have to be taught through educational programs and education at home. More attention should be paid to children's education, how do you expect to get loving adults, when they are not taught to behave that way during childhood? Parents play an important example role and will therefore have to show how we in the spiritual society can interact with each other in a nice manner.

This means that responsible parents should have sufficient time to spend with their children during the week. In contrast to the current situation, the number of hours that people have to work will be limited to make sure that there is enough time left for themselves and their children.

Giving love to children

Small children are the most beautiful living beings on Earth, aren't they? A child playing in the sun that looks at you with a friendly and happy face. Such a child is carefree and has a natural innocence. Children want to be happy and want to have fun. When you give love to your kids, they will flourish; you will bring out the best in them, and then you will receive their love in return. As far as a child is concerned a warm and happy family is a gift from heaven. He or she can feel safe there and grow up to be a wonderful person. Children also know, by nature, that love is the most important thing. In that respect it is just as if children are very close to the universal entity. As if they have just been there, in a world of endless love. There is nothing bad in a new born child. He or she radiates love and happiness. And often they stay that way through the first few years of life. As long as they cannot talk, they communicate mainly through love. They want to be held and hugged. If there is not enough love they pick up on this immediately. It is as if they want to shelter from the rain. They are pulling back and the smile disappears from their faces. It is therefore important that you as parents show that you love each other and that you love your kids unconditionally. This will ensure they will grow up to be healthy adults, both physically and mentally. This process of growing up is very important for our society and should therefore be well managed. It is a wonderful job to perform, although it is not easy. We cannot continue to fall back on: "But we have to work!". That was in an ancient primitive time. We now know what is important for our society. It would be logical that we make time for it? We should accept the consequences of setting the right priorities. Moreover, we have sufficient resources to meet our primary needs. That cannot be an excuse. So let us look at a well thought out plan to achieve this.

Taking responsibility

Before people have children, they should follow a mandatory educational program, which will teach them what a child needs and what in particular you should pay attention to. I think very few people know that in the first year of life a child needs to bond emotionally with at least one adult, preferably the mother. This is not in line with taking children to a nursery a couple of weeks after they have been born. On the one hand we give an important piece of responsibility away and on the other hand we increase the risk of negative psychological impact on the subsequent development of many of these children. In the spiritual society the need for childcare will be minimized. And where this is inevitable it should be done by pedagogically trained staff. In this way we prevent children developing emotions in an insufficient way and the continuous intake of healthy adults in society is secured.

Having children contains a certain responsibility that only ends after the children have moved into their own home and have become 18-21 years of age. This responsibility means something, something other than shifting this to others or giving children money and a key under the guise of 'entertain yourself'. This means only managing a situation, but certainly not educating.

Educational training program
What is that education going to look like?

The situation will not be much different from today, but we are finally starting to live up to our responsibilities. How do we do that? Here are some basic rules, which could lead to a better education, appropriate to the spiritual society:

1. **A mandatory training / workshop for upcoming parents:**
during the training the development of the child is explained. A child goes through different stages. It is made clear at each stage as to what the child needs and how to react best. In addition, the do's and don'ts are taught. Think of it as support for upcoming parents. Of course everyone thinks he knows how to educate children from past experience and with the help of familiar 'experts', like friends and relatives, but they do not always receive the right advice. With a little help, the first years of education can be a lot better for the child, creating a better basis to grow up to be a balanced adult.

2. **Both parents are working less after birth:**
both the father and the mother are required to work less during the years that the children do not go to school. One of the partners also has the opportunity to temporarily stop working completely, while the other is working full time. Not everyone has the same skills and needs. It is therefore better to give people some room to make their own choice. You can only educate children if you make some time for it. Working takes a lot of energy and people are usually tired afterwards. Most people do not usually want to do a lot of necessary things when they get home. This causes them to be irritated quicker than when they are relaxed. For many working people taking care of children in a positive way with a lot of attention is a problem. By giving people time to do

this, without any further impact on their personal situation, it will be a pleasant experience for many and the children's' education will benefit.

3. **Compulsory coaching by an education expert**:
a coach will visit each family regularly and give advice on (possible) problems until the child has reached its eighteenth year of life. The coaches may also offer additional training and facilities to be able to prevent serious situations.

The coach is not a controller, but a consultant, who helps people to keep the education of their child going in the right direction. It will be a form of assistance, where you can ask questions and talk with about your doubts, so that everyone can stay in control themselves. Often parents have questions, but have doubts about going somewhere to talk about it. The threshold is not too high. The threshold is much lower when you can talk to someone on a regular basis in confidence in your own home. Most of the time the talk will be about simple things, like what a child already can or cannot do at a certain age. But sometimes there are more difficult subjects, such as when your child must be tested for dyslexia.

4. **Make sure the home and the neighborhood are pleasant**:
a society is a society when people do activities together. This begins at home, where you must try to make things 'sociable'. This has little to do with buying a big house or putting nice things in it, but merely by how you interact with each other and the activities you do together.

If there is never anyone at home, there is no sociability. Sociability is created by people. When children come home from school, they want to tell their stories to someone. They prefer to tell their parents. The parents also have the chance in this way to keep in contact with their children and to know what has happened to them and what keeps them occupied. The child must experience home as a foothold from which he or she can explore the world. A cup of tea and a biscuit is often enough to give them the strength to go outside to play.

The neighborhood is often perceived as unsafe. The people of the neighborhood may change this, because if there were more people at home to keep an eye on the children, the parents would dare to let their children play in the neighborhood again. In a more open society neighbors know each other and can make agreements together on how to ensure that their children can play safely in the neighborhood. Police surveillance does not have to be a necessity.

The neighborhood is the first step for children in the whole wide world. Make sure that this is a pleasant and safe place to be!

5. **Teach your children, about why we are here on Earth:**
the basic education for the young adults starts at home. If you want your children to participate in the society, you will need to teach them, why we are here on Earth and how our society works.

The parents are the most influential people in the first years of their children's lives. They have the ability to provide them with a basis from which their children approach the world. Give them a gun and they will be soldiers or criminals. Give them a first aid kit and they want to be doctors. It is the same with ideas. Teach your kids the basics of our society and there is a high probability that they are going to behave accordingly. This is not only a task for the schools, but also the parents. It is important that you as parents set a good example. You cannot expect children to behave differently from yourself. Children will copy behavior of influential people in their surroundings. In most cases the parents are high on their list of most influential people in their environment, whether they show good or bad behavior.

By telling children how the universe works and why we are here on Earth, a basis will be created from which they continue to build their future and are able to grow as healthy and worthwhile people in the spiritual society.

6. **Teach them honesty, respect and love for all life on Earth:**
we all want to be treated with fairness, respect and love. The best way to take care of this is by teaching this to your children, by showing a good example yourself. This sounds simple, but give it a try!

Coming from an evolution of hunting, violence and wars, it is difficult to break this cycle. This can only happen if we continue to insist and are alert on changing behavior, indicating relapse into primitive behavior. Boys are happy when they have a fight now and then and like to be the boss. Cooperation is harder than it seems. All children have their own character and can clash regularly. If we ourselves are aware of this and take it into account, we can respond to it in the right way. We must also remember what we get in return from this investment, because who would not like to be treated with love in his old age? You cannot expect your children to look after you in your old age, if you have not spent much time with them. The most important thing is that we teach our children how we want to interact with each other.

7. **Teach them to listen to older people**:
Where young people excel by power and speed, the elderly have gathered wisdom and life experience. The elderly are on the other hand physically less able. They often need a little help. Teach your kids to listen to the stories of these people and to learn from them. The youth of today are the elderly of tomorrow. By listening, you can learn a lot and you are better prepared for the future.

Every older person knows: the time that you are young is very short. In the course of your life the way you look at your life changes. Whether it is the result of life experience or simply of the passing of years I am not entirely certain, but this is not important. The essence is that you can learn something from older people, because they have a different view on things. When you are young you usually look forward, but you have got little experience to base your actions on. Halfway through your life, you take a time out to look at what you have done so far and based on your experience you make plans for the years you hope you still have to go. When you are old and your time is nearly up, you start to look back on the past period, a period where you have gained a lot of life experience and wisdom that you should really transfer. It is a great pity if no one listens and all the knowledge of the elderly gets lost along the way. This contributes to the fact that we are badly prepared for life and at the same time we exclude a large proportion of the population. If we, as a society, want to move faster, then we must learn to listen to the elderly and to use their wisdom to our advantage.

8. **Be forgiving**:
Nobody is perfect, and everyone makes mistakes. That goes for you, your partner and your children. You do not have to agree with everything that someone else does wrong, but you must learn to handle these situations in the right way. Please do not get irritated by the mistakes of others, after all everyone makes mistakes. Help each other to prevent mistakes in the future.

If you want to learn to interact with people in a nice way, you must learn to forgive. This also applies to children. Parents should not condemn every mistake. It is important to teach your kids to go on after an error and that people do not always judge them by their errors. If we would take every error in account, then people would not be able to live together, no matter what kind of society.

Many disputes arise because people do not really forgive each other's mistakes. They remember all the errors, until a certain threshold is reached. The quarrel erupts and often leads to a situation in which they no longer want to meet each other. People who are forgiving can usually handle a lot of different people, while people who do not have this gift go from one quarrel to the next.

9. **Accept your child the way he or she is**:
We say that every person is unique. Many people do not know what this means exactly. In most companies, we try to classify people and forget the fact that every person is different. Talk to different people and let them say something about themselves. You might find out where that uniqueness comes from. Every child is also unique, therefore a standard approach has a limited chance of success.

An approach that works with one child can be a complete failure with another child. A child has its own character from birth, which may be totally different from another child even though they have the same parents. While one child may need space to flourish, the other may just be in need of structure and clarity. This means that first you need to figure out what kind of child you have been given, before you can determine how to educate your child. If you do not accept the character of your child and start to tackle it in the wrong way, the child will perceive this as a straitjacket. You are actually abusing your child. The final results may be far from what you wanted to achieve, a child who has never been happy or worse still, a child who is mentally scarred by his upbringing.

Therefore accept your child as he / she is, surround him or her with love and once you have figured out your child's character, you are able to determine the most effective way to educate them.

10. **Pay attention to your relationship with your partner**:
Another important factor for the education of your child: your relationship with your partner! The damage to children, caused by divorce of the parents, is enormous and highly underestimated. If you can prevent this, you have already done many good things for your child. Therefore time needs to be spent on your child as well as your partner.

Many children are traumatized by a divorce between their parents. In a lot of cases it is apparent in their behavior when they are young. With older children, we think that puberty is causing their strange behavior, and as they mature, we think they have out grown it. Nothing is further from the truth. Because of the copying-behavior of children, following the example of their parents, these people often make the same mistakes as their parents, often followed by a divorce in their own relationship. We want to stop this development in the spiritual society. Children need a good example and a stable base to grow up. This means that parents should keep working on their relationship and develop together. People must try to prevent growing in directions that do not work well together. Ultimately, this will always have a positive influence on the development of the child. And society will also benefit from healthy and happy people. In case of problems not only the family is confronted with these problems, but also the people around them as well as in the workplace. Ensuring healthy relationships is in the interest of different levels of society and therefore something to focus attention on.

More points to focus on

What a list, you may think, but this is just the beginning, because during the education of your children you will find out yourself that there are many more points to pay attention to. Most of them are really focusing points and not laws or regulations. In the spiritual society, we appreciate each other's help and we also find it important that our children receive a good education.

A coach can play an important role, because this person is more experienced with all these concerns. Therefore it would be a comforting thought that you can get help when you need it. This coach can also give support to the relationship of the parents. Educating is not something you do with two fingers in your nose at the same time. Hopefully it is clear that education takes time and that time must be created.

Less regulation

When people are taught during their education from childhood that too much of anything is not healthy, that balance is important and we must ensure that everyone is happy, this will lead to different behavior of people in society. This is essential, because otherwise you get a system with lots of rules. This restrains people in their freedom and gives them a sense of oppression. This is something we want to get rid of. We must therefore teach people to make conscious and responsible choices. Healthy people with a well planned education, who have learned to take account of the world and environment in which they live, and make healthy choices. These people do not need much regulation. This creates more pleasant people and a better society.

Custom Education
People are now being raised from the perspective that they have to survive in a capitalist or a communist system. Depending on the system where you grow up, it requires different skills and behavioral characteristics. If we change the system, it will also be necessary to change the way of education accordingly. Different output requires different input, sounds logical? We want people to behave differently in a spiritual society.

5.7 Moneyless system

The biggest change will be in fact that there is no money in the spiritual society. This transition cannot be implemented overnight. A society that functions without money is only possible, if everybody is well prepared.

Originally products were traded against other products. With products you could 'buy' other products. As a result of the need for more complex transactions and the division of labor, money came into the world. Especially in international trade, money appeared to be a good tool.

Also people who worked on the production of goods received money with which they could buy all kinds of products.

The value of products is zero
We have arrived at a point in time where we no longer want and need to pay for the production of goods. That is why we move every form of production to low wage countries in the current capitalist society. Many products are therefore of very little value, while the people who produce them are usually living in poverty. This could not have been the intention when we invented money. In addition, we see that where countries are in financial trouble or stock market threatens to plummet, debts are forgiven, funds are being supplemented or extra money is created. Money lost its usefulness as a trading tool long ago. Money and power are used in today's society as a justification of why one person has more assets than the other, and to be able to have influence on other people. And this situation is deliberately maintained. But this is not right. We say we want to give everyone the best or do we not? Then why do we not give everyone all the best? The simplest solution is to devaluate all resources to zero

and to eliminate money from the society as a whole. The link between money and power is then also gone, back to the essence.

Motivation from within

In the spiritual society everyone gets the same reward, as long as he or she is contributing to this society. But actually it is not about the reward. Your motivation comes from within. From the universal thought you have learned that it is important to develop yourself and to help your fellow man wherever you can. We ensure that everyone can live at a high level of prosperity and well-being. A level in which everyone can be satisfied, but also balanced with the world around us.

Contribution to society

It is important that you can prove, as a living unit or family that you have contributed to society by working a minimum number of hours per week or month. This is saved in a central database and everyone gets an ID-card[3], which you can use when you want 'to buy' something. It is of no use to try to steal this card, because the owner can immediately have it blocked and send in an application for a new one on his or her personal number. The number of hours worked is not transferable and becomes personalized. When you buy something, the number of hours worked is not lowered in the database. This number stays the same. The fact that things have a value of zero makes them unattractive to steal. The only reason to steal something might be because you do not want to work, but this is soon discovered in the spiritual society. A quick check in the database delivers the evidence. It can hardly be simpler.

When you go somewhere, the central database is contacted and it checks whether you have made sufficient contributions. And what if you have not worked enough hours? Then you get a few weeks to catch up on those hours. When you do not succeed you get a visit by a public official to discuss the problems and find a solution. Note: in this system it is not about punishing people, but helping people.

Central registration of worked hours

This central registration could, for example, be done by the current banks, since they have the facilities to make such ID-cards and make connections

---

3 An alternative to this identity card might be a microchip implanted just under the skin.

with all facilities, such as shops and institutions, where payment machines are used to perform checks.

Such banks are a lot simpler than the existing banks, because only a single registration is required. There is no interest, no debt, no stock market, no capital markets and no money being earned 'dishonestly'. This is too bad for countries such as Luxembourg, Switzerland and the Cayman Islands, but this system is just too simple for tricks. Of course it is important that the integrity of this system is monitored, but banking institutions are good at that.

Financial systems are no longer needed

All the systems that focus on money alone, that are not production based, are no longer needed and will be dismantled. Production means the provision of services or the making of products in order to serve people. The systems of interest and shares, as well as the tax-system will be dismantled. Insurance policies are no longer necessary, because everyone is always insured by the government departments. All these make the society complex, and they add nothing to the welfare of the people, on the contrary they are a burden. We are not going to do things any longer for money, but for the welfare of the people. Is this not motivation enough?

Benefits

And for whom this is still not enough; let me name a few advantages: we are no longer going to put people into small houses and cars. Everyone has a beautiful spacious house and a good comfortable car, so envy and jealousy are unnecessary. And you get enough time to enjoy it with your family. Everyone can go on a holiday three times a year. Poverty does not exist anymore. Crime has no use any longer, because there is no money! No more robberies in the street. No more intrusions into your home. Charities are also of no use anymore: we will just do it! We are really going to help the people in need. The society becomes much simpler. We can focus our energy on things which are really important, such as the treatment of serious diseases[4] and the development of environmentally friendly energy sources and production methods, and of course to help the weaker members of society and people in developing countries.

4 The pharmaceutical industry continues to exist, not as a commercial industry, but as a supporting sector for the society. Even pharmaceutical drugs that are unprofitable can be placed on the market.

5.8 Vision and policy

The current society

governments in the current capitalist society do not work with an integrated policy and the growth of the economy often seems to be an end in itself. They have completely forgotten that the result of their policy is meant to help the people. Who do they mean by the people? The shareholders, multinationals and investors or the ordinary people? Maybe they have just forgotten to help the last group.

The new society

The most important change, that must take place to manage the problems outlined in the previous chapters, is to replace the existing social systems by a new social system. Money plays no role (or is of secondary importance). The motivation to participate in this society should mainly come from within. People might see the government of such a society as the center of power. Therefore the recognition of the universal entity is the basis for this society. In this way, a new system is created that I would like to name the spiritual society. In this system all the power is granted to a universal entity (this makes power a more or less meaningless concept). There is no head of state, but only an elected parliament and government. They are responsible for the governing of the society. This form of governing is extensively described in the next paragraph.

Becoming one with the universe

The spiritual aspect is characterized by becoming one, which is the universe and everything within. In addition, a consciousness of the functioning of the universe should occur. Everything is connected together, life and death. What you do here on Earth, contributes to your own actual being in the universe. Your relationship with your environment is of great importance. This awareness coupled with the new social system, will bring out the good in all the people. In this way a balanced life will come here on Earth, where happiness is within everyone's reach.

Objectives

A society as a whole must be focused on something, to prevent different parts from starting to work in opposite directions. Therefore the spiritual society has to consider the following objectives:

Long term goal:

Survival of mankind.

This objective consists of three sub-objectives:
- Balanced society of people, animals and plants on Earth;
- Spiritual[5], mental[6] and intellectual development of all people;
- Establishing a human society within another solar system.

The concept of "balanced society" is mainly concerned by taking into account all living things, to ensure that humans, animals and plants can co-exist, the maintenance of ecosystems and wise handling of resources in relation to the surrounding world.

Looking for another living planet is at least necessary for mankind to survive in the long term, for example, after the end of life of the sun or in case of a (natural) disaster, making life on Earth impossible (for a long time).

Short-term goal:

Happiness, love and health for all people on Earth.

Preconditions:
1. people need to have a state of mind, which can be achieved through their upbringing and education;
2. a society should ensure that the basic needs of people are met.

Care and love for each other

The care and love for others are the key issues in the new society. What you have done and what you mean to others become important factors. The reason for this is that care and love for the ego comes naturally with humans. That does not mean that we must eliminate ourselves at the expense of others, because we must, of course, also take good care of ourselves. Otherwise we are

5 Development of the ego-transcending awareness, awareness of (everyone's place in) the universe, getting in touch with the people and the world around us on the basis of inner experience.

6 Development of the emotional, psychological and social side of the ego.

not able to help others. Each individual is equally important.

The training, performance and reward system is tuned to this. Knowledge and behavior are taught and maintained from these systems. This is much more of a system based on logic and rationality. Opposite to a so-called self-regulating system of which is suggested that such a system is fair, while in reality it is not.

It is possible for everybody to have a rich and happy life, whether you are a plumber or a doctor.

Freedom

This life is much freer than under the current system, because everyone has a lot more opportunities in his or her life. Now only the rich and powerful can do whatever they want. In the spiritual society, there is freedom for everyone. Nowadays our freedom is limited by our financial possibilities. In the spiritual society, the freedom is mainly bound by our shared sense of fairness and the need to balance our lives with the world we live in. No artificial border, but a natural one. Without this understanding and knowledge this new world will also be perceived as limiting freedom, because the natural limits are equal to everyone, without exception. And if you do understand, you will not experience these natural boundaries as limiting freedom, but as protection of our environment that we want to maintain.

A society of the next generation

The spiritual society is not a form of a communist or a capitalist system, but a society of the next generation. Think of it as a deliberate choice for a better society, one step forward. The good sides of both systems are maintained and there is also an extra something added to it that makes the spiritual society future-oriented: for the Earth and for all who are living on it.

5.9 Government and power

Democratic governance

As previously described, the spiritual society knows a form of governing in which all the power is granted to a universal entity (this makes power a meaningless concept more or less). But the voice of the people is also being heard and counted for. The description of the universe and the previously described basic rules are the basis for the government of the spiritual society. It should not be the intention to ever make changes to these basic rules.

The government is democratically elected, so that the government is supported by the people. This is just the same as in every modern democratic society.

The big difference is in the starting points, which are much less ambiguous and straightforward than in the current situation. Also, the policy is clearer and more focused, because we know what we are working for and why we are doing what we are.

Royalty and inheritance

The formal power over a country like the Netherlands, will no longer be with the royal family. They have a purely ceremonial role, at least if the people still want that. We should not forget that the monarchy is a legacy from the past, in which power belonged to the strongest force. We want to get rid of this principle. Inheritance is something that no longer exists in the spiritual society. Small personal items such as dishes, jewelry and photos may be passed on, but assets have no value anymore. You have got assets, because you need them to live. They are a tool, but not an end in itself anymore. Everyone has one house, no money and a company is led by a capable person. This company may be led by the son or the daughter of the previous person, who led the company. That does not matter. The power has become independent of the assets in this manner. There is only one official power and that is the universal entity. In addition, our population has become much more multicultural and these people have nothing to do with the monarchy. A major point in favor of the monarchy is the fact that it is a symbol of unity, especially to the outside world. A monarchy can keep the people together and can also account for some public relations abroad. The royal family must agree with the new basic rules of the spiritual society, because otherwise it will not work.

The government

The government is formed by a council of wise people. These educated men and women have earned their stripes in the society. A minimum age of around 50 years would be reasonable. These people should have the knowledge and experience of what life means and should be in touch with the people. A maximum age of 70 years would be right, because a large proportion of the population over 70 years is beginning to get weary and illness is a possibility. A wise person can thus be a member of the government for a maximum period of 20 years. Unlike the current government, the government of the spiritual society will not be replaced every four years. When a member has reached the agreed maximum age, he is replaced by a successor. This new wise person is chosen by the parliament, together with the other members of the government. A member of government can be someone from parliament, but it can also be someone from a political party or someone straight from the public. The choice is based on knowledge, vision of society and specific sectors. It is important that there is not only a short-term policy, but also an integrated long-term policy. The activities of the members of the government are reviewed annually by the parliament to protect the integrity of the government and their policy and to make sure they still have the right attitude to be able to remain seated. To be a member of the government is not a position of power, but a position of honor. The government is meant to guide the society. It is the responsibility of the government to develop a plan for several years to come in the interest of the people. Once a member of government is unable to do this, he will also need to recognize that a replacement is necessary. In consultation with this wise person another appropriate position will be sought. One possibility is that the former member of the government continues his work as a coach and consultant for the replacement member of the government.

Investments

The current government is often restricted in its actions because of a lack of money. Therefore they must often use measures (e.g. mileage) and laws (tax laws) to handle this situation. In addition, there are many rules to handle the (dis)behavior of people. This requires a lot of time and attention from the government. This is a big difference with the government of the spiritual society. In the spiritual system there is no need for all this, so much more attention can be given to the way in which the needs of people can be looked after. A lot more attention can also be paid to education, research and development. As a result, we, as a wealthy and innovative country, have many more opportunities to help others. The current government is far too involved in the controlling of the financial budget, thus bringing the interests of large groups of people in jeopardy. There is insufficient money from the government for research into causes and remedies for diseases. Institutes and foundations exist primarily on the basis of donations. Generally there is money for research into cancer, heart disease and aids, because this involves large groups of the population, but for diseases that affect smaller groups there is usually little or no money.

Also we all find the education in schools important, but in the current society there are insufficient funds. The result is that in districts with many wealthy people, there is enough money for good teachers and teaching materials from donations, while in the poorer districts this is not the case. The chances are so much lower for these children. In the spiritual society the level of education is the same everywhere, because money does not play a significant role and education is in the hands of the government. There is also a lot more opportunity of achieving the same high level of education at every school.

There is no money in the spiritual society. Therefore decisions on 'investments' are based on logic and the general policies of the spiritual society. On the needs of people, no one excluded.

Legislation

Part of the responsibility of the government is the legislation and defense system. The government will have to decide which laws, given the development of society, are necessary. Efforts will be made to keep the number of laws as low as possible, because less control is necessary. As children get more and more educated, this will actually be the case. Laws only have meaning, when they are understood, known and respected. All kinds of complex laws, which almost no one understands or obeys, are of no use. And if they are understood and known, but there is no control, then they have no meaning either. Because people do know right from wrong, there will be fewer laws and rules needed, but the control will be more intense, as there are no limitations in the form of money. This means that in the spiritual society people know where they stand: violation of laws will almost always be signaled and reacted upon.

Defense department

The state of the defense is determined by the situation in the world. I come back to this subject in another section. The government is, just as in the current situation, responsible for the deployment of the army. The army will however, be given a security function during construction activities in underdeveloped but stable areas, far more than they have today. The government decides where these activities are carried out. Furthermore the government decides on the development of the army, both in size and armament, but also about the tasks of the army. A spiritual society should have a defensive force at least, equipped with conventional weapons, strong enough to resist small-scale attacks of any aggressor and to be able to make an aggressor understand that an attack will not be without loss on their own side. Overall, this will be sufficient to keep small groups of aggressors out of the country.

Parliament

The parliament is elected once every 5 years by the political parties. Political parties will still remain, because the people will continue to show some kind of differentiation and thus will have different opinions. For example, race, religion, residence (urban and rural) and needs (the elderly have greater need for health care, while young families want facilities for children). In this way, the society stays focused on all relevant issues in the interest of all the people. The main task of the parliament is on the one hand to keep in touch with the population and on the other hand to assess the policies of the government and then put forward proposals to solve bottlenecks and issues. When they disagree with the government and there is no solution, a government member may be relieved of his or her duties and the parliament may designate a new member of the government.

The parliament is elected once every 5 years by the political parties. The political parties, propose potential candidates for the parliament. The people can vote for them, and the political parties will decide who may represent the people in parliament. It is actually not much different from today. It is important to note that democracy is maintained.

Because money is no restriction, many more wishes can be fulfilled, leading to a greater satisfaction among the population. It will be hardly necessary to make compromises. It is more about defining the optimum situation and thereafter to achieve that situation. Consider the following example: if there are traffic jams, build more highways or expand public transport or do both. Money is no restriction, so why would it not be possible? To create a situation in which the people are no longer able to go to work is not a solution. Think in terms of positive solutions, solutions that make people happy and joyful. Not just a few people, but everyone, it can really happen!

5.10 Government departments

In this system it may be clear that there is a major role for the government departments. Occasionally it will be necessary to take action to keep the system running smoothly.

Tasks

Control and monitoring will be an important task to identify excesses in the society and to transfer them to the appropriate government authorities. But even more importantly, the government departments will be there to inform, educate and help people to find their way in society. This relates to the idea that people cannot be controlled by rules which come from the outside, but that this should come from within. Their own rules should match the outside rules of society. The foundation is laid by education. As a result the government departments do not need to have a strong regulatory function. They can be more of a service delivering, supportive organization.

Despite a good education, there will be people who are losing track as a result of an escalating argument or conflict, a mental disorder or a physical malice, which may be caused by or due to sickness or an accident. This may be revealed in different ways, maybe first signaled by the police. The government departments are there to arrange institutions taking care of such situations. People who are in trouble should have some coaching to try and put them on the right path, if this is not an option this person should be placed in a section of society where he or she feels at home and cannot harm anyone. It makes no sense to let people, with permanent mental defects which make them a danger to the lives of other people, stay in society, and you run the risk of further casualties, which must be avoided.

Responsibilities

The government authorities will be held responsible for all the services that are offered to the participants in the society. Think of the energy- and water supply, health care, food supply and public transport. They should make arrangements for these kinds of general services that provide the basic needs, so that there is no need to worry about it. A national board should be installed. This will not be much different from the current situation. They also should be monitored to prevent them from becoming a dominant organization. As stated above control and monitoring are tasks of the government departments. The monitoring function should be separated at the right level from the more executive functions of the government departments and may be directly under control of the government itself. Anyway, these are all fine details. This should be completed when it is time to do so.

At this point I would like to stress again that the major difference lies in giving people equal chances and opportunities for people, creating more freedom and a greater chance for a happy life for all. The provision of free basic services, such as housing, transport, food and water, energy, clothing, education and health care will contribute to this. The existing institutions, which now regulate these facilities continue to exist, but must report to the government authorities. As long as they can demonstrate that they are running their business well, there is no problem. The operation of the system is roughly the same. Citizens must make agreements with these institutions about the supply of products and services, but if insufficient response has been given on repetitive complaints the government authorities are required to intervene.

Role of 'bank' for companies

Regarding the allocation of space and buildings for companies, the government authorities play a major role. I will come back to this in the paragraph on companies.

In the present situation, people often start a business because they just want to give it a try or because they think they have no other option. There are a lot of them that go bankrupt after a while. In the spiritual society, the Chamber of Commerce and the government authorities have a role to lower the number of companies that go bankrupt. This does not mean that fewer companies will be launched, but the government authorities have a more active role in helping businesses start in the right place and to create the right situation for a company to be successful. In this way we can prevent many people from ending up in negative situations. A financial bankruptcy no longer exists in the spiritual society. This reduces all the problems that people are confronted with nowadays when their business comes to a standstill. On the other hand the government authorities will keep an eye on whether there are sufficient sales within a company to go on with this business. They will agree in advance as to how many sales should be generated to be able to call it a successful business. On the basis of the stock records, the government authorities will do some supervision. If the government office than decides that a company should be eliminated, there will be a discussion with the 'owner' to find a solution. You will receive help to search for another job, or to find other opportunities to run a business. The starting point is positivism: look at the potential of people and their positive attitude. Then there is always a possible solution that will work in the long term. People make mistakes, learn from them, and come out stronger and more successful so that they can handle anything.

Informing people

Another important task is to keep the population informed, so that we can continue to act responsibly towards other people and the environment. We stop making products, which form a threat to the human health. Our health is not something to play with. Everything we do is done for the people! Isn't it illogical to do something that endangers people? In general, for every problem there is a solution. In the current situation there is money involved in anything that carries risks, but once money is no longer a factor, there is no reason to do so. Nuclear power and diesel cars, they will no longer exist, as well as SUV's, which pose a danger to pedestrians and other road users. These kinds of cars have higher energy consumption than necessary, caused by their heavy weight and poor guidance of the airflow. In addition products that cause radiation will be eliminated from the world. Not to mention genetically modified foods, hormones in food and chemically raised livestock. The government authorities have a duty to inform the people and make them aware of the risks of such products. In addition, it will also ensure that such products are no longer made or imported.

Evaluating and adjusting

The authorities of the government department are not almighty, because they know what their role is and how the universe works, and they realize that love is the main focus in this society. Furthermore, we will need to evaluate people's thoughts on the government departments, their rules and ways of working on a regular basis. The government departments must develop according to the country's situation. The government still has a role to ensure this happens. The government departments must be accountable to the government, and the government will make the necessary changes to the appropriate departments to keep them functioning in an efficient manner.

And what could be better than an institution that takes care of all these simple things? Life contains enough issues for which it is much more important to pay attention to. Or not? How many people have not outsourced their household and care for their children? Then let us not worry about government authorities who are primarily concerned with much less important matters?

5.11 Legal system

The legal system is focused on what is fair, responsible and honest, and all other points do not matter. The spiritual system is a simple and straightforward system. The rules and laws which have remained are fair, topical and simple for anyone to understand. In addition this prevents extensive and complex lawsuits. Crime is already limited to a minimum, making it much easier for perpetrators to be identified. Lawyers are no longer there to keep offenders from prison, but to examine whether someone is innocent or not, and if someone is proven guilty the lawyer's are there to help find an appropriate punishment. It is in nobody's interest, when a killer or Mafioso comes back into society or worse, is acquitted. This method increases the likelihood of good citizens reporting as witnesses. This means that on violation of laws and rules there will be tougher penalties, then in the current system. Now we still use money to pay off the punishment, but that is totally useless when criminals have a lot of money. People do not learn by these punishments and it is certainly not a solution. Punishment should be executed shortly after the act of crime, because it increases the chance that it has a psychological effect on the offender, this should be manageable due to the simplified legal system.

Disputes

The legal system will be much less called upon, since there are fewer rules and laws in the spiritual society, than we have nowadays. What undoubtedly will still regularly happen are disputes. People will have different views, which they cannot agree on, causing disputes to arise. The legal system will then assess what is fair and reasonable and come forward with a solution or supply people with new insights, so that they can hopefully come together. This way of dealing with cases is meant to find out the truth and to give people the feeling that justice is being done in an honest manner. It is certainly not meant for outsmarting each other.

Prisons

Despite the fact that we are starting a new harmonious society, this will certainly not be the case in the beginning, and incidents may continue to occur. In those kinds of cases we need a legal system that helps people to keep on track. Prisons, as we already know them, will remain a necessity. They will, however have to focus on reintegration of the prisoners, more so than the current day. Once these people are released, they need to have a future, to prevent them from falling back into a criminal situation or environment. This only applies as previously described to those prisoners of whom the probability of relapse is almost zero.

Chance of recurrence or irreparable suffering

People, who cannot be helped or have killed someone intentionally, should not return to society. The suffering that they have caused is irreparable, or the chance of recurrence of the offense is too big. These people will be brought to a place where they can live, but cannot return to our society. With the help of the many communications we have today, they can keep in contact with their closest relatives and friends. The risk that they will make more victims is unacceptable for a society.

Juvenile delinquency

When it comes to children, the responsible parents will also have to be involved much more. It is all too easy for parents to exempt their custody, and then they do not learn anything. Parents and children will be held accountable together and be addressed together. In the spiritual society this will come down to a mandatory therapy to get education back on the right track, this therapy can be given in their own environment, but in more severe cases it can also mean that the whole family will be brought to a special location for a longer period of time, where they can be observed properly. In short: dealing with problems quickly, effectively, starting right from the core. The whole society will be served by this approach.

Police

The justice department is responsible for the police force. The people will obey and have more respect for the police in the spiritual society. Because there are fewer laws and regulations, the police can focus on helping people, with things like accidents and major events, and the managing of movements in society, such as the traffic on the road and waterways.

More district police offices will also be established, where people can come 24 hours a day. These police offices will be located in the center of the district, usually in a mall and/or neighborhood center. The local police officer is a central point for the people in the neighborhood who will be able to help and give advice on various problems and bottlenecks.

Monitoring the government authorities

The legal system also carries out checks on the functioning of the government authorities. Reports of regular checks will be sent to the central government and the government authorities itself. The government authorities should start by trying to correct the situation. When a situation of repeated inaccuracies and/or irregularities occurs, justice will notify the appropriate government authorities and finally ask the government to intervene. Mismanagement and corruption are thus avoided. This has to do with the principle that everyone needs to be monitored. It keeps people sharp and prevents people from doing things wrong. This applies to the lower staff as well as managers, because these people have shown in the past that they are not able to provide the right example. And it often comes down to the right example: a good example will be followed without a doubt.

5.12 Products en facilities

Equilibrium

The basic principle of the new economic system is balance. We continue our production process, but this time in balance with nature. Economic growth is not necessary, because with a rising prosperity the population will decrease (see paragraph 2.12). We are therefore almost certain to have a constant population. Money will be abolished. You need not fear that no one is going to work. During the education of children they are taught that they should contribute to society throughout their lives to develop themselves and even after their death they can go on with what they have learned from life. The new people are looking at another way at life. In addition, work has become fun. Companies are societies in themselves, where the welfare of employees is an important issue. The disappearance of the shareholders will undoubtedly contribute to this fact.

Free take away
Why should we pay for facilities, which everyone needs, and of which we agree that everyone should have? Think along the lines of food, fresh water, energy, a roof over your head, a car, public transport, health care etc. Let people choose a house or apartment with a maximum of 500 square meters land or 200 square meters floor. Everyone still has the right to a roof over his head? Let them get food and drinks, when they need them. You do not want anyone to starve, do you? You cannot eat and drink more than is necessary. Agreements can be made for one or two cars per household. And so on. Why should we need money for these things?

Registration of purchases
Anyone can buy whatever he or she wants. With your ID-card, you can go to a store and get the items that you need. To prevent your things from being stolen by someone who does not work, all the products will be provided with indelible encodings. When you buy the products, they will be registered directly to your name in the central database. If someone has products, but has not worked for them, this is very easy to clarify.

Scarce products
What do we do with products and services, which are not readily available for everyone, or not available within a short time? A good example is when the whole world wants to go on a holiday to the Maldives at the same time. This is simply not possible. To begin with, the whole world will not be in the same situation at the same time, as described in this chapter. It will take a lot of time and so we have time to deal with these changes. In addition, not everyone has the same needs. This is partly determined by the family situation, age and health. In short, the chance that everyone wants the same at the same time is very small. I personally would not like to go to the United States, because so many people carry a weapon or to China, because it is a dictatorship, where people are oppressed. Or to Africa, since I would not like to arrive in a slum or hospital. I can mention a lot of places where I would not like to go. My parents are about seventy years of age and they do not want to leave the country anymore. However, it may happen that too many people want the same. But that is of course the same today too. Even rich people must sometimes wait years for their orders, because, for example, a special limited edition Ferrari has to be hand built by a

small team. In the new situation, people will sometimes have to wait for their orders. By tuning the production into the needs of the people, this problem will become smaller over the course of time. And for a trip to the Maldives you will go on a waiting list. Otherwise, go to another tropical island. If the world population is not growing beyond all limits, everyone's individual needs can be met. If there are too many people, this may be a problem, but then this is not the only one. In that respect it is also an awareness problem. With reasonable and logical thinking, we should be able to solve all of this and become one large happy family.

Responsible production methods and products

The purchase and production of products is constrained by a system of standards and values. Cars with a fast acceleration and cars that form a danger in any other way to pedestrians and cyclists will no longer be produced. This is the same for cars with abnormal energy consumption, which burden the environment too much and the same goes for cars that produce substances, like soot dust, which can cause cancer. Aggressive-looking cars will also no longer be built, but of course, there will be no more demand for such cars. People will think more about the relationship between their actions, purchases and to their surroundings, the world we live in.

Furthermore, the production is determined by what products people want to purchase. This is based on things like environmental friendliness, comfort and ergonomics and of course what something looks like, than on what people can pay. An additional advantage is that paying people, often children, very little money for their work, which has happened in low-wage and developing countries, is over. Producing under appalling conditions in factories that do not meet current western environmental requirements will be no longer necessary. The usage of illegal workers and immigrants does not happen anymore, because only legal workers, those who are registered in the central database, can have their working hours registered. Illegal workers cannot. Owners of companies still exist, but the legal situation is completely different. A company cannot be sold. Moreover, these owners earn the same as their employees. They do not earn more by pushing their workers to the limit and they cannot give their employees very low wages. The exploiting of people no longer exists. There are no more people who can enrich themselves on having a business of their own. It is also not necessary, because everyone is 'rich'[7].

We are developing many products that are not required in the current society, salesmen hope that people will buy them, to generate money. That is turning the world upside down! To make objects, because you want to earn money and not because people need them. These types of products usually disappear after a short time. The really useful products, the commodities, are the products which we need to survive, to travel and many products to communicate. You might wonder whether we need much more.

What have we learnt about this in the spiritual society? We are careful with the raw materials of the Earth. We produce in an environmentally responsible way and only those products that are in demand. In this way, there is little waste material. This does not mean that all products are the same or as sober as possible, but we are aware of what we are doing and how we do it. What people do not want or need will stay in the shops. Nowadays people buy these things, only because they are cheap. Good research before production is very important. Creating a new product is a real process, which needs to be thought about in advance!

Companies will have the responsibility to provide safe and comfortable products, and everyone will get the chance to buy them. This means choosing a responsible road in the middle by both parties. Cars are a good example in this case: an SUV is safe and comfortable, but environmentally unfriendly and unsafe for pedestrians. A typical middle class car is also safe and comfortable, but a lot more friendly to the environment.

Complete facilities

The same applies to many facilities. In many cases, money is the threshold for making facilities more comfortable. This starts with rooms in retirement homes, but it can also be seen in the interiors of holiday accommodations. It is easy to go on and on describing examples. A very clear example comes from the insurance industry, where they offer health insurances, which give a 50% refund on different kinds of treatments and 75% on others. But when I need

7 In the spiritual society 'rich' means that people are happy, having a high level of welfare. In terms of prosperity that they can have anything to improve their quality of life, without showing excessive behaviour.

an artificial leg, what can I do with only half a leg? The other half should really come. In short, this is offering something which is just not good enough.

Money, however, was only intended for bargaining, but nowadays it has become the yardstick for all and a limiting factor that does not make us happy. By making money disappear we are able to create facilities for the people, which make a real difference. We can offer people something that they have been waiting for.

5.13 Economic system

Traditional economies are based on growth: more sales, more profit. There are more and more hands needed and thus more and more people. This leads to pushing employees to the limit until they have had enough, which in turn leads to population growth. This causes many health problems and overcrowding, too many people in one place leads to all kinds of problems in itself. A few examples: traffic jams, people who jostle each other in order to get a seat on a train, lack of clean drinking water, lack of energy, intolerance and lack of houses. Many of the problems in our society are caused by over-population, encouraged by our economic system.

Equilibrium

The economy in the spiritual society is based on balance, in all respects. The economy does not need to grow, because we rely on a balanced composition of the number of people in our society. In addition, companies have no shareholders, so there is no need for higher profits. Indeed, there is no money, so how would you be able to express profit?

The economy is based on the principle of living in balance with the Earth. The companies, for example, may only produce waste that can be recycled or which the Earth can process. That means that we will have to think carefully about what and how to produce.

Quality and ergonomics

Another advantage of an economy that is not based on profit is that we are able to pay more attention to issues such as quality and ergonomics. Why would we continue to produce products that no longer meet the required quality standards? The price is no longer a reason to buy a product of less quality. In addition, people will only demand products, which they really need. So no more small cars or furniture that is terribly ugly. These things can be taken out of production.

Competition

Competition is another principle that has got a completely different meaning in the spiritual society. In the capitalist system this means performing better than someone else. This is reflected mainly in producing and selling more than other employees. Once a company is making losses, they run the risk of going bankrupt, which causes significant problems for the owners. The employees will lose their job and income. In the spiritual society it can still happen that one store is selling more products than any other. The other store(s) will then have to determine what is causing this, it is not about realizing the maximum profit, but selling products that people need. The shop owner will have to go on searching for products with better marketing opportunities. If he fails, a shop will not go bankrupt, but he will be assisted in finding solutions (See paragraph 5.10 and 5.14).

In short, there are many challenges even in this economy, but they are more focused on the people involved (e.g. managers, employees and customers) and not on the money. Not on realizing more, but on balance, realizing the same is good enough.

5.14 Companies

The business owner

There will still be companies, but there will no longer be shareholders. The central government department will determine where certain companies are needed. You can also specify that you want to start a business somewhere. If you have the right qualifications, and have successfully completed an assessment[8], you will be regarded as a prospective director/business owner. Subsequently, the government authorities will determine whether the company that you want to initiate is achievable and realistic. To take over an existing business is therefore a little easier. This company has already proved to be worthwhile.

8 A test to determine what kind of business a person is able to manage. E.g. with or without staff, a service or production company etc.

The state does not own companies. The owner[9] is the person who runs the business. The government authorities are qualified to assign someone to a company, to shut a company down or to end a business. The state has, in a way, taken over the role of the commercial banks. Incidentally, this is a slightly different role. Whoever owns the company is no longer important in the new situation. The owner can therefore give the company any name he wants, even his own name: what's in a name? The only thing the government department does is to prevent things from going wrong. Once permission is granted to start a business, it is up to the director/owner to make the business work and to ensure that it is a well-run company (similar to franchise?). His success will be primarily determined on the one hand by whether he knows how to create a business friendly environment for his employees and on the other hand whether he knows how to sell enough products, which people need, and whether his company is customer friendly and grants services. Success will no longer play a role, so the success will not be determined by how much profit someone has made, but by the things that really matter: has this business added value to society, for the people?

Staff

A director or manager must ensure that the company is still running well and that the personnel have satisfaction in their work, and that they are able to deliver quality. What if the staff are not doing a good job? Then the manager has to talk to that employee about what is happening. If the problem cannot be solved over time a coach has to be called in, with who the employee can look for a solution. Maybe there is a labor dispute or there is a need for some different work. Pushing people to the limit is no longer needed. The amount of staff should be adequate for the workload: not too much and not too little. So no more: "more and more work with less people!" This is an important task for the new managers.

The company as a society

By changing the way in which companies work, they will become societies in themselves. People will no longer feel pushed to the limit. They will go to

9 The concept of ownership in the spiritual society has a different meaning than the current one. An owner cannot sell, trade or give away his company in the spiritual society. You can give something back, causing the process of determining a new owner to start again from scratch.

work feeling comfortable in their working environment. Absenteeism will decrease. Employees are more likely to give their opinions and preferences. This allows for more involvement. Employees will be more willing to participate in company-activities and to search creatively for new solutions concerning the production process. Employees in many companies will also get more opportunities to attend courses. This does not need to be training courses focused on their current job, it may also be more soft training skills, allowing an employee to develop himself further as a human being. In this way the manager has a major task in dealing with the human aspect. A manager is more of a coach than a boss.

Regulation

When a company does not sell much, it might be better to shut it down. The director / owner of a new company will be granted a few years to build a sound business. Just like in the current situation. Thus there is a chance that some companies will disappear, because they are no longer viable. The government department could very well delegate this role to the Chamber of Commerce. An experienced director / owner has priority in the creation or acquisition of a further business, if he or she to.

Another reason for a company to shut down is because it cannot get the staff. It is therefore the task of a company to make the work enjoyable. People will really start looking for a nice job, but companies who make the jobs attractive, are more likely to get well educated, smart and hard working employees.

Business transfer

The situation is slightly different than in the current situation, when it comes to business transfer. Nowadays a lot of companies are transferred from father to son or sold to a foreign company. This is not the case in the spiritual society. When a father wants his son or daughter to take over, this is only possible if he or she passes the assessment. Then the son or daughter has the first right to take over the company. If not, then the company will be transferred to someone else. Everyone will be taken care of in the spiritual society. Therefore there is no reason to claim possession.

A separate item is formed by the farmers companies. There are also large pieces of land attached to these companies. These will not be split, if there is no need. A successful and well-run company should be changed drastically.

When there is a good reason to redistribute the land, the reasons and the way in which it should be done will be discussed with the farmers. Because the standard of living of these people is guaranteed, there should be no problem in solving this situation.

5.15 Work

Registration of worked hours

People are rewarded for the work they do and in particular for their contribution to society. The hours worked are administrated in a central database. The value of the working hours is the same for everyone, regardless of one's job. Registration will take place from the moment that someone is 18 years of age. Until that age, children are dependent on their parents and all large purchases should be done by them. Children receive a child ID-card which lasts until their eighteenth birthday, with which they can identify themselves and purchase for example food and drinks in limited quantities. If someone is still studying after his eighteenth birthday, an annual dispensation can be obtained. The dispensation is determined by a designated committee.

Work distribution

For families with children, for example, the number of working hours lies around 60 hours per week (for 2 adults). For singles and couples without children it is 40 hours per person per week. The number of holidays is about 40 days per person per year. Depending on the age and their health condition variation is possible. One big difference is that almost all the work in the spiritual society is seen as real work, including volunteer work and work for social clubs and associations. When you want to work there fulltime, go ahead. The value of the working hours is equal to any other job in society. The point is that you do something that gives you satisfaction and of which the government authorities have determined that the work contributes to society. This will apply to many more jobs than is the case in today's society.

Fun in the workplace

The work that you choose, is something that you like to do. When you want to do something else after half a year or so, then that is possible. This is obviously a bit complicated for the professions, where much training is needed or where there are certain physical requirements. In any case there is no reason not to do something else. The earnings stay the same. People will therefore be happier in their work.

Research has shown that people are the happiest when they do voluntary work. And the more they are doing voluntary work, the happier they become. Please note: we are talking about work! And it makes people happy! This shows that there is no need for a material reward to make people work. When the basic needs of people are satisfied, people are willing to work without any financial reward.

It is nonsense to think that certain professions will no longer be exercised or that people will not work hard enough. In the spiritual society, people are aware that they do it for themselves and each other. Moreover, people do their work because they want to do that work, not because they have no other choice. And if they want something else, they are just going to do something else. Thus you have more motivated staff in this society.

Special achievements

Sometimes people suggest that extreme and special achievements are only made, when there is a special and high reward. But then they have forgotten that the people who really do heavy work, such as miners and stone cutters, generally receive low wages. The people, who have earned extreme amounts of money, usually have done relatively little for it. This inequality does not exist in the spiritual society: people do high level sports, because they like to compete with other people, not because they get well paid for. Athletes are again drinking beer together after a game and are not pulled away by their sponsors to a press conference to proclaim and portray the fact that he could have only done this with the help of his sponsor or some other statement in the interest of his current sponsor.

Do not think that employees will take it easy. People really do know from each other who is working hard and who is not. This is simply identified and discussed with that individual. If there has been no change observed within a few weeks, a coach will be called in. Together they will examine the cause and if necessary find another job. But if we really start working on educating people and people are aware of their place in the universe, they will be willing to make something of their lives. Ultimately, everyone wants to be appreciated, to receive and give love.

Personal development

People are working in companies, because it is fun to work there, because it contributes to the society, because it is a way of getting into contact with other people. And if you want something else or need to have more variety, this will be arranged without affecting your personal life. The amount of hours you work will be valued the same, regardless of the work you do. It only needs to be work that clearly contributes[10] to the society. In this way, people have more opportunities to look for jobs they like or that suit their personal development, because personal development is one of the principles of the spiritual society. We are on Earth to help each other and develop ourselves. This is done by gaining experience in life. This includes the work situations that you encounter in your life, but also the training that you have followed. In this way, learning and working can be a very good incentive for your own personal development. You cannot expect that the work someone does on his twentieth year of age still applies to that person on his sixtieth year of age. But if that is the case anyway, it is likely that that person has developed very little in the meantime. By doing new things, you will be inspired to develop yourself, to learn new skills, you will get new insights. It broadens your horizon. This is a logical and accepted way of living in the spiritual society; therefore everyone will be supported during his working life to be able to make the right steps. If someone wants to change jobs, it is important that professional support is provided. Choosing a future job and/or training is easier when you have help from a qualified person. With this support in the spiritual society and the lower risk of consequences for such a step, this will be easier and less stressful. A career path will become a natural and logical thing to do and not something you are forced to do, as is often the case with management trainees.

Retirement

Everyone will work until their 65th birthday. If the work is becoming physically too heavy or is for any other reason not suitable anymore, they need to search for another job. The living conditions of that person will remain the same, causing less stress, and there is no increased risk of becoming ill. The work will be more of a social function and will be a place where people feel at home.

---

10 As a community we determine what contributes enough and what does not. Nowadays we only pay for the services and products, which are worth paying for. This is the same in the new society. Products and services, which are of no use, will not be accepted. They will automatically disappear.

It is also a fact that not everyone will be able to deliver the same work performance until their 65th birthday. If there is a need for an earlier phase-out scenario or a different job, with another company for example, that is an option. This will always be in consultation and with the consent of the concerned person.

Stress

Stress will barely be an issue anymore, work atmospheres will be healthier, because nobody wants to be subjected to such situations any longer. We all know that suffering from too much stress for too long is not healthy. The work will be nicer and healthier. Who does not want that?

5.16 Health care

The people who work in health care are doing so, because they want to help sick people. Because there is no money at stake in this system, there is no chance of care becoming priceless. More than enough doctors can be trained and employed, as budgets and salaries are not relevant. Each hospital is independent and regulates its own affairs. However, there is responsibility to the government authorities. When there are things that a hospital is not able to arrange, they can turn to the government authorities to take care of it. In addition, the government department has the responsibility of ensuring there are enough care centers and well-trained staff. Enough space needs to be present in hospitals and nursing homes to help everyone who needs care. Care is one of the basic services and should therefore not be left to individual initiative. It is a responsibility of the government department. Care must be accessible to everyone. We will need less care, because society is healthier and more responsible. As a result fewer people will get sick.

In the hospitals the atmosphere will be better and less stressful, because nurses and doctors have more time for their patients. In short, better health care is available for everyone!

Research into diseases

A final, and important point, concerns the study of diseases. A lot of that research will be done by universities in the spiritual society. The duties of a university concern primarily education and in addition research. By making the universities responsible for research, they can have the most modern equipment for research and students will be allowed to become acquainted with them. There will be more opportunities to grow within a university. These universities will be linked with production companies that deal with the production of medicines and medical equipment. The lines between research and production should be reduced and the tasks of universities made broader.

Patients are people
Care for people is not a business activity, but a human activity. Care is given to people who are not fully independent and who are often in pain. It is difficult when people are no longer independent. The care of these people should be carried out with love, not because of the revenue. In the spiritual society, we help each other, because we like to see everyone happy and we want to have meaning for each other, not because of the profit we can have from one another. This is another example of another way of thinking. Motivation should come from within the people and not from the outside.

5.17 Agriculture and stock-breeding

Production of healthy food
Agriculture will be more focused on the welfare and health of people, than it is today, it is no longer about mass production, but about producing healthy products. They have to be produced with responsibility. This means that plants and animals will not be stimulated to grow unnecessarily. No genetic manipulation or the use of hormones. Animals should be given normal food, such as grass and water and no wastes from other processes.

When mass production has ceased, diseases among animals and problems with manure waste will be significantly reduced. In addition the waste process will also become more balanced. Balance is the key.

No additional flavors and colors will be added to food, just to make it look and taste better. These substances can have negative consequences for people, while they do not have a necessary function. Something similar applies to preservatives.

Production close to home
The production needs to take place close to the selling point. This makes it almost unnecessary to use preservatives. It also saves unnecessary energy that will occur during transport.

An additional problem caused when transporting food is that we are dependent on other countries. Regularly we are confronted with producing countries where the human rights are not respected, production is done under appalling conditions or the environment is being polluted. This is not only true for the food industry. We block these issues out, because we are dependent upon these countries. If we decided to create a fuss, we run the risk of the food chain shutting down; it is actually an unacceptable and hypocritical situation. By retrieving the production of the most essential foods to our own country, we solve this problem and we can look our fellow man in the eye again and say it as it is.

Agricultural employees

Today, agricultural employees, the former farm servants, have become priceless, as a result of the living standards in most Western countries. Many illegal immigrants and seasonal workers are being deployed from abroad during the harvest season. This causes all sorts of difficult situations, including the housing of these workers.

An agricultural employee is rewarded equally to any other in the spiritual society and thus he can live in a normal house and lead a normal life. The life of the farmers becomes much easier and people, who like to work on a farm for a while or like to do seasonal work, get the chance to do so.

Responsible import and export

After the introduction of communism in Russia under Stalin, collective farms were founded. The land was being taken from the farmers and a large portion of the food was destined for export. Ultimately, however, too much of the food was being exported, causing the farmers to starve. The government held the money in their own pockets. Let me be clear: we do not take anything from the people and the food supply of our own people is the first priority in the spiritual society. All things should be considered from the standpoint: how can I keep the people happy and healthy? And when you educate people in this way, you will see that everyone is ensuring that this really happens and intervention is not necessary.

That does not mean that we are able to provide everything ourselves. And on the other hand it is good for contact to trade with other countries, but is important that we do not need to do this for the essential products. If there is a reason to stop trading with another country, this should be possible without causing any major implications for the population. Otherwise, there is an unacceptable dependency. A major problem could arise in delivering, not only as a result of conflicts between countries, but also for example due to a natural disaster.

Packaging materials

Packaging materials should be reduced to a minimum or should be reusable. Otherwise this costs unnecessary energy and burdens the environment. Why are beverages such as soft drinks, milk, alcohol, etc not sold in disposable containers and in large reusable cans? That would save us a lot of packaging. In several regions of the world, heaps of garbage are already causing enormous problems. In addition, the burning of waste causes a lot of pollution.

Developing third world countries

Large parts of the world hardly have any good stock-breeding and agricultural areas, while this situation is unnecessary, if they put some of the energy into the construction of irrigation channels to help dry areas. By making these areas greener, this will ultimately make them less dry. If we do not fight this development, eventually all the Earth will dry out everywhere and the entire globe will start to look like the planet Mars. To me, this seems like a good argument to take some action, don't you think?

Nowadays it is all a matter of money. Without money, we can take action and by means of a rotating system these kinds of jobs can be divided among several people. By ensuring that more parts of the world are able to provide food for their own people, fewer unnecessary deaths will occur in these parts of the world. Also fewer people will move away from areas where they have lived for generations. Usually these people are attracted by the cities, where more and more slums are developing, because these people tend to lack sufficient education. By giving people the chance of a normal life in more places in a country, the population will be divided over the country, thus solving multiple problems!

5.18 Energy

Environmentally friendly energy production

The energy of the future is extracted from natural sources, with the environment taken into account. The current possibilities are limited, however we largely have ourselves to thank for this. As long as there are sufficient fossil fuels at a reasonable price, there is almost no one willing to invest (again a money issue!) in alternative energy sources. The shareholders must be kept happy! And if it does happen, it will be because of the fact that people are afraid of their image, which might have an impact on long term business profits.

Dealing more consciously with energy

The use of energy can be significantly reduced by many different kinds of measures. We often use more energy than necessary. More often we are not even aware of the fact that we are wasting energy: making virtually unlimited phone calls, using navigation systems (when we usually know the way), chatting on the computer (with friends who live nearby), driving in circles, and flying to the shops. The use of energy can be reduced strongly by using energy in a clever and responsible way. Once again, this is an awareness and educational process. As long as we do not recognize the value of indirect contributions to society, we are falling behind and we will always be confronted with problems we could have avoided.

When the world population is balanced and production is designed in a responsible way, energy consumption does not need to take extreme forms and there will be enough for everyone. The environment will not be burdened excessively which could have caused huge changes. Changes will always be there and natural disasters cannot always be prevented. If money is no longer playing a significant role and the welfare of the people is the key, we will be there to protect people from natural disasters and make appropriate arrangements to do so.

What about solar panels and windmills on the roofs of houses? If we combine these with a really good insulation in homes, and take into account the position of the sun, the hot and cold side of a house, many homes would probably become self sufficient in their energy needs. It is also an awareness process: do we want to share the energy with others or do we demand everything for ourselves? In the spiritual society, we take each other and the Earth into account. We are aware that we are not just here for ourselves, but also for each other. By saving energy here, someone can use this energy elsewhere, and vice versa. Do not think that it is only you being considerate to others. In the spiritual society, we are constantly thinking about how we can make a difference to those around us. It is a way of thinking that is unbelievable in the current society, but one that has a future. It is only in this way that we can save ourselves and humanity. In a system where the starting point is more, more, more and where people should compete with each other, this way of thinking will never work. Restrictions will be quickly perceived as oppression and in fact, that is true, because some people are preventing others from reaching the same level of prosperity, while they have the right according to the same system.

Alternative energy production

We do not want to produce energy from fossil fuels and nuclear energy anymore, but from wind, water and solar energy. And what about the heat from the Earth? Look how much energy is released by a volcanic eruption. The Earth bubbles with energy within, but we don't do anything with it. We have not made enough progress in this so far. We pretend to be capable of a lot, but at the same time we are spending a lot of energy on useless things like gadgets and motorsports, etc.

The new energy sources are producing clean energy. We have not pushed ourselves to the limit yet to get the maximum out of these energy sources. These are, however, the energy sources of which we can explain the use to our children, without having to feel guilty about the negative effects.

Energy Companies

The current energy companies may continue to exist, but will be once again subjected to government supervision. Energy supply is a far too important facility to leave to the individual or foreign initiative. Energy is one of the basic needs of people and therefore people must be able to rely on having energy, no matter what. Outsourcing is out of the question. A number of rules will be drawn up by the government and companies will have to comply. The energy supply must be guaranteed and produced in an environmentally friendly manner. Furthermore, these companies have the freedom and the support of the government to investigate more efficient and environmentally friendly energy sources. All from the perspective of searching for what is best for the people and the Earth.

5.19 Water supply

Water is one of the main sources of life for humans. It is therefore very important that clean drinking water is available for everyone. The Netherlands is one of the leading countries in having the knowledge to produce clean drinking water. On the other hand, we are also wasting a lot of drinking water by using it to shower and for washing cars. We are aware of such inefficient processes in the spiritual society. It takes a lot to purify water. We therefore decide to use purified water only for drinking. We will also produce simpler purifying plants for water for other use, such as dishwashing, car washing and showering. We do not need this water to drink and therefore it can be cleaned in a much easier and less energy consuming way.

Transfer knowledge of purifying water

The knowledge, that a country like the Netherlands has gained about the purification of water, will be transferred to other countries around the world. This enables them to build the same purification installations, so that people everywhere can have healthy drinking water. Water is the prerequisite to be able to build a life. Therefore we will have to educate other countries as well to learn how to deal with the available water efficiently, so that the current dry countries can prosper. Wouldn't it be nice if all over the world, flowers bloom and people can sit under a tree in the shade? It is very well possible, but we should help these countries with the knowledge and facilities.

Responsibility of the government department

Water is so important, that will fall directly under the responsibility of the government department. The same goes for the energy supply. The government department may ask companies to take care of the actual supplies, but the government department will be addressed, when there are problems with the water. In any case, the policy towards the water is a government matter, which cannot be outsourced. The government should consider the measures, both short and long term, to be able to continue providing everyone with clean drinking water.

5.20 Spatial planning

How do we organize our lives? The spiritual society will change all this. People will make other choices, because money is no longer a restriction. Many people in apartments and simple homes will leave for nicer homes. Also, many people will go to the countryside. Many more people will want to work and live in a natural environment. This has a huge impact on spatial planning. The government authorities and project developers have always played a highly restrictive role in this matter. That role will probably remain, but freedom is the new magic word.

Give each other room

Many simple houses and flats will have to be demolished and room will have to be made for beautiful detached houses with parks and playgrounds and lakes. The cities should be more livable. Many small towns will have to accept large numbers of new inhabitants. This can only work if it is done in a responsible way. Everyone will have to be aware of the fact that change is there and that we must give each other room. You cannot just take and expect that others will just give. Sometimes you need to give and at other times you can take. We will have to teach people that nothing is more beautiful than giving in the spiritual society. This will make the problem of spatial planning a lot easier. All parties will have to take each other into account of course, but when everyone is willing to do this, acceptable changes can be made.

Creative solutions

We must also be creative occasionally in this situation. Sometimes there will be no other alternative than to build a highway or railway in someone's backyard. In the spiritual society people will understand that sometimes there is no other possible solution. People are more willing to work on a solution thanks to the reduced attachment. This could mean, for example, that a new home needs to be found for these people, where they can feel at home. If these people are forced to move, it is only fair that they receive preferential treatment in finding a new home.

Another aspect is the living space. The Netherlands is a densely populated country, but in the north of the country there is still plenty of space. For financial reasons, this area is under developed. When more people move into the area, we should use it more efficiently. For example, suppose the number of immigrants is growing fast and we want to accept all these new people? Why can't we create a new area as big as the Netherlands in the North Sea? There is enough space. It is just a matter of how creative you want to be. Money is not the limiting factor.

5.21 Traffic and transportation

Decrease of home-work traffic

Traffic flows will always remain. But in the spiritual society the traffic from home to work and vice versa will decrease significantly, because a lot of people will find work for themselves in the area where they live or will try to create work there. People do not want to depend on a few large multinational companies, which are far away from their homes. Employment will therefore spread across the country. This will also ensure that people are not tied together in one place, but are living in all parts of the country. Many people do not want to live in large cities, where everyone is living and working close together. The result is less traffic congestion.

In addition, many will seek alternative modes of transportation such as the bicycle, motorcycle and scooter, because people are living closer to their work and there is also a growing awareness of environmental issues. The car will no longer be a status symbol. This is partly due to the fact that everyone can buy the same cars, the question of whether one needs a car and if it is really necessary is becoming an important factor, because it creates added pressure on the environment.

Extending the traffic network and public transport

The current traffic congestion problems will soon be resolved, even if the distribution of the work across the country is not easily achieved. This will happen in the first place because growth is not a target anymore and second because money no longer plays a role, which takes away the obstacle of developing the traffic network further. This may be by widening the existing highways, or building roads above the existing roads. Although they must not meet at the end because then there will be congestion at these points. As an alternative, for example, the traffic should be led to large parking towers around the cities. From there the people are transported with fast public transportation to the center or industrial districts. What about floating trams or cable cars and fast elevators? There are plenty of options, only a little creativity is needed.

Freight transport

With regard to freight, we will need to search for more environmentally friendly solutions to be able to transport large quantities of cargo. Nowadays that is done with huge polluting diesel trucks. We should make more use of water and airways to work this out. In the Netherlands for example a railway for freight transport from Rotterdam to Germany has recently been built. In addition a huge transfer harbor would be built near the German border to enable the exchange of cargo between ships and trains. As a result a lot of jobs would also be created in that area. This harbor never came and so now we have an inefficient railway. Another solution would be the use of zeppelins. These can transport a significant amount of cargo on a relatively low amount of fuel over long distances without suffering delay from traffic lights or traffic congestions.

It is also possible to build special transfer points around the cities from where the goods are brought into the cities by environmentally friendly smaller means of transport.

Conclusion

It is really not so difficult to think of solutions, which are agreeable for everyone. In the spiritual society people know what it is all about: living life in a responsible manner and trying to keep each other happy. The absence of power and money will already solve many issues.

There are plenty of opportunities to fix existing problems, without doing this at the expense of certain groups in the society. We can really achieve this in the spiritual society, so that we can go to the beach on Sunday together without having to stand in a traffic jam for half a day.

5.22 Foreign affairs

The main task of the Ministry of Foreign Affairs is to show that the spiritual society is the ultimate form of living together and maintaining friendly relations with other countries. It will also take care of making agreements with other countries.

Intermediary role

The spiritual society may need natural resources and nutrients, which are not available on its own territory. In that case, trade with foreign countries is necessary. Of course it is possible to trade with a spiritual country, although in principle such a country uses no money. In the instance that the other country also has a spiritual society, there is no problem. Agreements on the mutual supply of raw materials and/or goods can be made on the basis of friendly relations. This is not about exchanging things of the same value, but to give something to each other because you want to. Trying to help each other as much as possible taking into account the limitations of one's own country.

In the case of a country with a traditional society, we need an international currency to be able to pay for the supplies. This issue is explained in more detail in the chapter, in which the migration plan is discussed. It is after all a temporary situation.

Foreign Affairs have the task of maintaining friendly relations with other spiritual and non-spiritual countries, and to ensure the establishment of the exchange of raw materials and goods. Foreign Affairs are concerned with making good agreements with these countries. You will understand that to provide as much as possible in one's own needs is essential for survival. It also makes life simpler and is burdening the environment less by having shorter supply lines.

Promotion of the spiritual society

The main task of foreign affairs will be to promote the values of our society. To show the other countries how our society works, so that more countries are being stimulated to transfer to this way of life. It is easier doing business with these countries, than those with a non-spiritual society. It also reduces the likelihood of wars in the world. And the more countries that are based on a spiritual society, the greater the chances of human survival.

5.23 Defense

The need for armed forces

As long as the spiritual society has not yet expanded to the entire planet and there are countries with a different system, it will be necessary to maintain armed forces to scare off possible attackers.

At the one end is a heavily armed army, equipped with nuclear weapons. The other extreme is doing without any army. I think we must realize that violence has never been a solution and never will be. Changes that have been created by violence have never been permanent. Violence is something negative, because it is associated with psychological and physical pain. As long as there are still major weapons in the world, an equivalent form of defense will be necessary. Through negotiations this needs to be reduced to acceptable proportions.

The equipment of the military

As long as there are imperialist countries, the army will be equipped with modern weapons. However, horrible weapons like nuclear bombs and chemical weapons, which can even cause victims a long time after use in a conflict, are not allowed. On the other hand the army will be equipped with special tools that allow them to continue fighting when such weapons are being used by an enemy. When a country is well trained and has equipped armed forces at its disposal, it will deter armies with nuclear weapons. Since the strength of the army no longer depends on the amount of money, this offers a lot of possibilities.

Self defense

The aim should be to have an army of martial arts experts based on a Buddhist principle: self defense. They should be armed with tranquilizer rifles at the most, to be able to make dangerous people and animals temporarily harmless from a safe distance. We must realize that a terrorist and a guerrilla fighter are likely to have family who are still alive and who are non-combatants. By killing people the process becomes irreversible and hatred will start to spread, which will turn against you in the long run. Violence does not bring peace; you bring peace by offering people a future.

The home country

The army also has to deal with conflicts within their home country, which the police are not able to control. In principle this should not be necessary, but there is always a possibility that a situation can get out of control, even if it is just a tiger that has escaped from a zoo or circus. In addition, security is a task of the army, for which deployment of police is not enough. The army will also be deployed to disasters in their own country, when many fast hands and equipment are needed. The army is a professional army, since special military skills and physical abilities are required to be a soldier. Soldiers over 40 years of age are transferred to other positions of their preference in society. The police and humanitarian aid are good options, but basically anything is possible. Where necessary, soldiers will be helped by allowing them to follow a retraining program to prepare for a new job.

Abroad
Outside its borders, the army will not be used for military purposes, only for dealing with humanitarian actions. This means that they will provide security in foreign countries to aid agencies, keeping them safe from danger while working. This also means that they will not be deployed to war zones, because the equipment is not sufficient for those kinds of situations. The most important factor in this case is of course the peaceful nature of the spiritual society. The spiritual society will stay neutral at all times during foreign conflicts. Other countries may join our society. When the latter happens, it may be considered to deploy a joint defense force and to help each other in case of an attack from the outside.

I am convinced that a peaceful attitude will convince other countries to choose the same kind of society in the end. Especially once they see the high level of prosperity and welfare of the people in the spiritual society and that nobody has to live in poverty.

5.24 Police

The police are of course also essential in the spiritual society. There will always be people who, for whatever reason, become derailed and are causing problems. The police are there to ensure that these people are being arrested, after which they will be judged by the justice department. The police forces are lightly armed for this task, they will carry such items as a baton, pepper spray and a gun which produces a tranquilizing electric shock. This will ensure fewer injuries and deaths as a result of police violence, and the police are also showing a good example by not using lethal weapons, but only resources to temporarily make opponents harmless.

In addition, the police officers will be training together with the soldiers in the field of martial arts. When a police officer has reached the age of forty, he no longer has to do field work. Other tasks should then be found either inside or outside of the police force. Retraining courses are available.

Furthermore, the police are responsible for prisons and the regulation of matters such as traffic, lost property, assistance in fires and accidents.

An important aspect in this case is that there is no restriction to deploy enough police officers. The expectation is on the other hand that by the different principles of the society and the different education of the people, rather less police will be needed than in today's society. In the spiritual society, there will be little or no crime, because there is no money, regulation is minimal and simple and everyone can have all the means available, provided that a contribution to society has been made according to the mutual agreements.

5.25 Entertainment

Entertainment and making people laugh is an important aspect of the current society. People need to relax from time to time. To watch people doing something funny, singing a song or performing a play, is a good thing. Lots of music has brought a lot of people enjoyment. Music represents a certain spirit of time and it is possible to express certain feelings or to denounce some wrong doing through music. It is also a form of expression for people, and they can really throw themselves into it. There must be opportunities to continue to do this. In addition, radio and television have an informative function, which is very important. People know what is happening in the world. This makes everything on Earth more transparent, and this should not change. However that there will be no more exorbitant salaries in return. Currently, many media personalities earn huge salaries, while this is by no means a good reflection of what they are producing. It is also a business that is often based on knowing the right people. And when we look at the behavior of these people in private, it is generally not to be used as an example.

They are very much overvalued. The cameraman has often had more education than the artist on the other side. It is a good thing that all people are equally rewarded. If you still want to be an artist, then that is your choice and you must do it, because you want to, but not because of the money.

Advertising will only be focused on informing people and not on capitalist principles. The power of the media will become much less. The nature of the entertainment will become a lot more responsible. Deliberately hurting people is no longer tolerated. Entertainers will be taught that they have a responsible job, because they reach out to large groups of people. Often they are an example for many. When an artist is giving a bad example, he or she will be relieved of his duties and replaced by someone else. The likelihood of such situations is very small, because everyone gets a better education and lives by better principles.

The operation of the entire entertainment industry will be very different from today. Nowadays it is still a question of big money. That will be no longer the case in the spiritual society. People who want to make a music-CD/DVD or who want to play in a movie will get a fair chance. There will be committees set up to assess who is the most qualified to play in a movie or to make a music-CD/DVD. It will also be possible to take part in TV-trials in selection programs during which the public may cast his vote.

A limited number of companies will be created to make movies, music-CD's and DVD's. They have a purely executive role. Their task is to take care of the quality aspect, after it has been determined what should be filmed or should be put on CD. It is important to ensure that this business does not fall back into family business or friend's politics. This tendency will be much less than it is today, because the position of the people in this business is equivalent to any other.

5.26 Tourism

Tourism is a major industry today. As already noted this should be no problem (see paragraph 5.12). Tourism will also remain an important industry in the spiritual society. People want new experiences, to relax, to be active for a few weeks in some nice place or just relax and do nothing in a different environment. It is a good thing that people are offered these opportunities.

Equal opportunities and chances

When you want to book a vacation in the current society, we know the system 'first come, first served'. Personally, I see no reason to change this. The main benefit of the elimination of the price aspect is that all accommodations should offer high standards. People will not be satisfied with less, except if that is the only accommodation available at that time or because they just once want a simple holiday. Everyone has the chance to travel in the same way and anjoy the same experiences. This means that everyone basically has the same choices and the same opportunities.

There is a slight chance that some holidays get overbooked. To prevent people from being excluded, it might be an idea to make a long term plan over several years. In this way everyone who wants to book such a trip will actually make that trip one day.

Financing foreign travel

Of course we want and we must continue to travel abroad. The government will ensure the necessary budget is available. Business travel will be claimed by the company. For tourism everyone gets a new credit card each year. This card has a financial ceiling and is only given to the adult population. It can only be used abroad and is being paid at the end of the year by the responsible government department. With this budget vacations can be made and things can be bought abroad.

Tourists who want to come to our country from a country with money need to ask for a temporary ID-card. For a fixed amount per week people can come to our country. The card is valid for a limited duration and can be used to buy food and drinks, to pay entrance fees to museums, amusement parks and to purchase items and use public transport. By creating this income, the foreign travel of the inhabitants of the spiritual society can be funded.

The less able

The group we must devote special attention to is the less able. They also want to have a nice day out, and experience new things. In the spiritual society this group is taken into account. They can earn the same rewards and therefore make the same kind of trips.

5.27 Sports

We now see that sports are being dominated by companies and the urge to perform for these companies, so that they receive even more sponsor money. This leads to doping use and other unhealthy situations. Athletes tend to extend their borders to be able to perform even better than ever before. And when they have reached their top and cannot perform better and start to lose, they sometimes seek their refuge in illegal drugs, which are delivered by all kinds of dubious people who just want to make money, no matter how. This is the wrong way.

Healthy top sport

Top sport will continue to exist in the spiritual society. Sport is entertainment and a social activity. The Olympic Games and many major events unite people and countries. This is a great thing and something that we want to maintain.

In addition athletes are an example for other people when they show us that delivering high performances can be fun, to try to get the best out of yourself and competing with others in order to be better than the rest. The athlete is the central point in sports, not the sponsor.

To accomplish this, sport will be dissociated from business, and athletes will be encouraged to deliver performances in a healthy way, without the need for doping. We need to recognize that it does not matter if the performances are not delivered.

The development of athletes

Children can become members of sports associations and take part in sporting competitions, including international ones. Once children have shown to have talent and are performing significantly better than the rest, they can choose to become a full-time athlete. This will be accompanied by a national organization, specialized in coaching athletes. The athletes will get a reward, which is equal to all other inhabitants of the spiritual society. Winning is for the honor and nothing else.

Once their performance declines, there will be arrangements made to make a transition to another social function. This is good for the development of the person and gives others the opportunity to become a full-time athlete.

Losing

By giving athletes the chance to be themselves again and by making their income independent of their performance, the sporting honor becomes dominant once again. It is nice to win an event, but you should also take the loser into account. In this respect athletes can be an example for the people in society, athletes who show that the loser is still important and is treated with respect, and that winning and losing are not important for your position in society.

It is important for people to see that there is room for everyone in society. You need to have some kind of internal motivation. Not from the outside as it is still in today's society, because otherwise you are left behind and receive no money to live off. It is at these times people start to do extreme things, arranging sports in this way should be avoided.

Healthy performances
By making athletes more equal and by giving them the same resources and the same reward, the use of doping will soon diminish. In this society there is no added value: why would you take the risk of destroying your body, when you get nothing in return? (Drugs can have all kinds of dangerous side effects).

5.28 The world population

Right to freedom of movement
The current world population gradually causes more and more problems, which you can also imagine on a smaller scale, when you do nothing about it. You may still be wondering, why? Look around you, there are so many people being born that everything starts to collide and people are starting to get agitated with the people around them. We are increasingly creating regulatory systems, such as mileage and other price-enhancing mechanisms to make it harder for people to go anywhere. This is antisocial and limiting freedom. Freedom is an important principle in the spiritual society. This also means the freedom to go wherever and whenever you want. People need to be free to move and to explore the world around them.

Being careful with what the Earth has to offer
On the other hand, when we consume more that the ecosystem of the planet can handle the system pops apart, causing a large part or perhaps the whole world population to die. Ultimately, this will happen anyway at the time the sun stops burning, but that is a few billion years away. However if we continue with what we are doing now, the world will have reached a critical point in a number of decades.

Instead of doing something about the growing world population, we are increasingly ensuring that the limited resources are available to fewer and fewer people, the people with money and power. They are already the only ones who can afford to visit the most beautiful places in the world. Isn't that something you wish that everyone could afford?

Slowing down the population growth

An important prerequisite for the survival of our ecosystem is regulation of the world's population. This does not mean closing the borders or to prohibit people from having children, but ensuring that a balanced society is being created.

You may already have noticed that in most western countries, where prosperity has reached a certain level and people are having more things on their minds than raising children, fewer children are being born and a more balanced situation arises regarding the population. In developing countries a lot more children are being born, because many children die young. In addition, it is also a development issue. Development of the prosperity of such countries is therefore of great importance.

On the other hand, when money and infant mortality are no longer great problems, you might see a sudden increase in the number of children per family. This can be avoided by giving these people information by educating them on the pedagogical aspects that are important to bring up a child in a healthy manner. In addition, getting too many children limits the development of one's own potential, another aspect that should also be taught within education. This will ensure for a stable population in the long term.

Certain religions are emphasizing that contraception is opposite to the written rules of the religion and extending the circle of faith is important. Faith is made for man, not man for the faith. Faith is a tool in life. Also everyone with a logical mind can easily imagine that a universal entity would prefer to see children being born in a loving and conscious way and not by accident, reluctantly or as a result of violence. It only makes sense to have children if we have something to offer them.

A spiritual society will ensure that there is a more balanced composition of the population, because the starting point is balance and not growth. This will prevent a lot of problems in the future. If the world population is stable and is no longer growing, there will be room for the animals and plants on Earth. The natural resources can be divided fairly and we do not need to crawl over each other like ants.

How nice it can be

We are always talking about the negative effects of overcrowding. But imagine how beautiful the world can be, when the world is no longer overcrowded, when there is room enough for everyone? If you can stand and walk wherever you want, without having to be afraid of colliding into someone else, without having to stand in a traffic jam while going to work? This can only happen if we all work together.

5.29 Emigration and immigration

Emigration and immigration are also issues that must be taken into account.

Emigration
Many people emigrate because they are not satisfied with the society in which they live. They think they are able to build a better life somewhere else. When the spiritual society is working as it is meant to be, there will be few people who want to leave. In order to reach out to people, who want to leave anyway, those people should be given a sum of money to start a life elsewhere if they want to go to a country that functions with money. The government authorities will know how to deal with this situation. The money that is needed comes, among others, from immigrants. They will have to pay a certain amount of money, when they want to participate in our society. For this payment they will receive an identity card with an identification chip. For emigrants who want to come back, the same applies. Dual passports will no longer exist.

Immigration
Rules for immigrants will have to be made to prevent too many people wanting to live in our society. Immigrants need to have worked here for at least one year in an open position, which demands some form of relevant training. The potential immigrant may temporarily stay in a house especially for immigrants. After a year, a review should be conducted to see if the expectations are met and that this person can continue to work here. It is also necessary to consider the absolute number of people to ensure that this does not rapidly increase. The total area of our country will remain about the same. If everything is all right the immigrant will be granted a permanent residence. This evaluation will be completed within a month. If the immigrant cannot stay here, the paid amount of money will be returned and a ticket provided to a destination the former immigrant wishes for.

The nature of the immigrants will change. In today's society immigrants are often attracted by money. Often they send money to their country of origin. The main reason immigrants will come to the spiritual society is to be part of this social community. There is no money anymore.

Illegal immigrants

To prevent people from staying here illegally, it is important that people are aware of their responsibilities. Illegal immigrants are maintained in the current system mainly because they are working for very little money. There are always dubious employers who want to make use of these people. In the spiritual society there is no money anymore and thus the existence of illegal immigrants will disappear. An illegal immigrant is in fact just as expensive as someone from our own country. In addition, illegal immigrants without a valid ID-card cannot obtain food and things, unless they receive them from others. Because everyone is always wearing his ID-card (you have to use this to pay for everything), people without an ID-card are quickly traced and removed from society.

Beggars and tramps

Beggars cannot be tolerated. They will be given help and will be transferred to a shelter home. People generally become beggars as a result of psychological problems or because they no longer feel at home in the society. In both cases, society has an obligation to bring these people back on the right track and to offer them a place where they do feel at home, abroad if necessary.

Asylum seekers

There will no longer be asylum seekers in the spiritual society. The current system has shown that this does not work. By bringing asylum seekers to the rich Western countries we do not know if they are genuine refugees from a brutal regime or that they only come for economic reasons. The costs of relief in the western countries are enormous.

Asylum seekers in the spiritual society will always be sent back to the region where they came from. Asylum seekers should find a connection with the regular flows of refugees from problem countries and will be given help in neighboring countries in the same region. The countries, which are hosting large numbers of refugees within their borders, should be provided with international aid to be able to house and feed the additional population members. We will need to travel much more ourselves to be able to help asylum seekers abroad instead of doing it in our own country. The starting point is always to return to the country of origin.

5.30 Welfare

What is the effect of the spiritual society on the welfare of people?

Happier than ever before

The spiritual society does more justice to the most basic needs of people and also to the things that most people in the world believe in. People in the spiritual society are having enough to eat and drink, a roof over their head that they may choose themselves, good health care, a feeling of freedom, a nice job, opportunities to develop yourself, to live in peace, a healthy environment, giving meaning to life and a society where the concept of 'togetherness' has a meaning from a sociological and spiritual point of view and not from an economical point of view. Actually, all the good things that you can imagine, without needing money and a great abundance. I believe that everyone can imagine a world like this. This would allow people in the spiritual society to lead a healthier and happier life, than the current one.

Equal opportunities and chances

The main causes of mass misery in the world, the culture of greed and overcrowding, are eliminated in the spiritual society. People have an equal chance for a healthy and happy life. In the current society this is primarily determined by where you are born, financial or social status, your parents, your physical and intellectual abilities and the contacts that you have. In the spiritual society none of this matters at all.

Everyone who works hard, who develops himself, who acts socially and contributes to the spiritual society, has equal opportunities. There are no reasons for you to feel disadvantaged or deficit being done. Everyone has the same income. Differences are therefore based on personal choices. It no longer matters what a person earns. Everyone's performance is valued purely on the basis of a sufficient contribution, without an exorbitant salary being paid. People are valued for who they are, what they do, not for how much they earn or for what assets they have. It is only about the really relevant matters in life: love and care for each other, and the development of yourself and humanity as a whole.

And what do we do with the disabled, sick and disadvantaged fellow human beings, you will wonder perhaps? Of course there is place for them. When you have read the section about religion and spirituality very well, you will understand that healthy people have a responsibility and obligation when it comes to the weaker members of society. This means not only a task in the form of care, but first and foremost to ensure that these people feel involved in the spiritual society. That they have an equal place where they feel valued. Where they are respected and where they can participate, they have the same rights and the opportunity to make similar choices. There is a place for everyone in the spiritual society.

More room to live

Economic and population growth are no longer the starting points and people are aware of the effects of overcrowding. Therefore, this will not be an issue in the spiritual society. In all respects people will have more space to live and develop, which will make them feel more at ease, happier and healthier. Traffic jams will be a thing of the past, as well as waiting lists in health care institutions and associations and supermarket queues.

Healthy people and relationships

By indicating the importance of education and development people will realize that having too many children is not doing anyone any good. That does not only go on the expense of the end result in education, but also on yourself and possibly your relationship. People in families will also start looking for a healthy balance. People will realize that more, more, more is counterproductive.

Families will become more harmonious and children will be calmer. Children will be able to develop better, which will make the differences between children smaller. This ensures a more balanced society, where people have a greater understanding of each other and are better at communicating. Due to the fact that people do not push each other anymore and because there is more personal space, there will be less irritation. As a result people will approach each other in a friendly manner. Each person will receive, as well as more living space, more room to develop themselves personally. The uniqueness of each human being has more room to flourish and will be more valued than in today's society.

Psychological and sociological welfare

It is therefore obvious that people in the spiritual society will feel more at home in psychological and sociological terms and will be happier. The friends you make are genuine friends and not just friends because you have got money or because they have a certain interest in you. People are not poor or inadequate if they have no money, these problems do not exist in the spiritual society. On the one hand this makes people more valued and gives people more equal opportunities, equal opportunities to a valuable life, equal opportunities for happiness, but also more opportunities to develop yourself in a direction that suits your own personality. It is more about how you handle your personal characteristics. How you behave towards other people. What you mean to other people. People have plenty of time for each other, because it is all about the people. It is no longer about the money, but for love for your fellow man, regardless of who that is.

Giving more meaning to life

Almost everyone believes in and hopes for a life after death, so why are we not living by that belief? We need to recognize the universal entity in general and hence the need to make something of this life, it will make the general meaning of life on Earth more meaningful. Not every individual will have the opportunity to search for the meaning in life. Some people because they die young, while others hardly have the chance to spend time on it, because they spend all their time trying to survive. People in the spiritual society are growing up with a meaning to life. This will give them a direction, which gives them the chance to develop further than many generations that have preceded them. Our children are therefore getting opportunities of which most of us could only have dreamt. I am not talking about money or materialism, but about spiritual wealth. Where that will take them, we can only guess. Maybe it will become very empty here on Earth as every soul on Earth succeeds in the ultimate development.

By starting to live after the essence of what it is all about: love, people will really start to feel involved with each other and the term 'society' will have meaning. Power is not important, there is only one universal entity and everything is connected within. That fact will give people peace and a basis for the generations to come.

5.31 Astronomy and Space travel

Limited lifetime of the sun

This issue is very special within the spiritual community. By now we know that the Earth will not last forever, because the sun will stop burning after a few billion years. The sun is a normal star and stars have a limited lifetime. At the end of its life as a star the sun will briefly shine a lot more intense and almost all the planets in our solar system will scorch, including the Earth. The astronomy is reasonably well capable of calculating the lifetime of a star. We will have to do something with that fact. In the current egocentric society many will say: it will last my time on Earth.

Survival of mankind

One of the objectives of the spiritual society is: survival of mankind. This is among others translated into the creation of a society in another solar system. Our sun will still shine for a few billion years, but it is wise to look for another planet within another solar system where the human race can continue to exist. In order to do so we will have to develop spaceships with which we can travel to such a planet. Currently, our space ships are not able to do so.

Development of space travel

In the sixties, we flew to the moon. After some walking around over there, Mars became our next destination. We have never landed there, only with unmanned spacecrafts. We also started to travel faster on Earth through the use of jet planes. Then the Concorde was discovered, a futuristic passenger aircraft. And soon we came to the conclusion that traveling in an airplane-like vehicle in space would be ideal. This resulted in the Space Shuttle in the eighties of the last century. However, in the meantime the Concorde has been retired to the museum, and the Space Shuttle is also in doubt, as a result of a series of accidents.

The biggest problem that we know is that the current jet and rocket engines are using a lot of fuel and oxygen, which is not available in space. Refueling in space is not an option. In addition, the speeds that can be achieved with rocket engines are not high enough to travel billions of light years to another solar system. Therefore we need new technology and perhaps even a new scientific theory should be developed in order to find a solution.

Unidentified flying objects (UFO's)
In the fifties and sixties many UFO-sightings were reported from all over the world. The number of reported observations greatly declined in later years. This is not surprising because in the decades that followed faster aircrafts, satellites and such became more common. As a result people became more accustomed to seeing things flying in the air.

What made the observations in the fifties and sixties so special is the fact that there are many documented observations from highly educated people with an understanding of aircrafts and space travel, such as pilots and astronauts, as well as observations by radar with which the speeds of the flying objects could be measured.

Research of the UFO-phenomenon
These kinds of observations have led to several official investigations of the United States government, such as Project Sign, Project Grudge, Project Blue Book and the Condon Report, which are the most famous examples. All of these investigations are from before 1970. And in all of them the conclusion was that the evidence found was not sufficient enough to justify further research. The result was that there has never been a more extensive research into the UFO phenomenon. Apparently, this is a closed chapter, with which you would make a fool of yourself if you wanted to open it again. [11]

A well-known scientific researcher of the UFO-phenomenon was Prof. Dr. J. Allen Hynek. In the seventies, he was head of the astronomy department of the North western University. During a period of twenty years, he has been consultant of the U.S. Air Force during the main investigations. He was initially very skeptical, but through sound scientific research, he came to the conclusion that there is more going on[12]. Unfortunately he did not succeed in a real breakthrough.

Many observations are unexplained to the present day. Statements from government authorities are hardly substantiated or are after a while withdrawn. It never comes to real serious investigations. Not even from the scientific side. What is going on, you might ask? Is this really not interesting enough or is this a cover up subject? Is the information too valuable to make

11 From the book "Visitors from Space" by Adolf Schneider 1976.
12 From the book "Live on other planets" by Chriet Titulaer 1976.

it public or does the government not want to disturb the general public? Or are these indeed covert operations from Area 51[13] with new experimental airplanes?

The journey continues ...
But back to why I have described astronomy and space travel here. For a long-term survival of humanity we will have to search for solutions outside the Earth. That is why we need the astronomy and space travel. Until now, the astronomy has not yet found a planet where we could live when our sun stops shining. The space travel has not yet provided us with the spacecraft, to be able to travel to other galaxies. I could easily have written a piece about the current state of affairs regarding astronomy and space travel, but that does not bring us where we need to go.

I believe that we can only go beyond the known borders if we are able to search along unorthodox paths and develop new scientific theories, with which we can travel in a different way and much further than before.

5.32 Education

Teaching the basics of life
All the previous comes together in the educational system. The raising of children forms, together with the education at school, the key to the success of the spiritual society. These are the places where the basic spiritual training takes place. Knowledge, understanding and

awareness of what love in our world and the universe can do. What the universal entity means to us. Spirituality is the basis for the functioning of the spiritual society. People get the right attitude to actually start realizing the spiritual society, when they have been taught about it by their parents and at school.

Education prepares people for the transition from the current society towards the spiritual society; in addition it also has a continuous role for children and adults to learn about the past, present and future. Education prepares them for their role in society. A good understanding of society is essential.

13 Area 51 is a top-secret military base of the U.S. Air Force and is located in the Nevada desert north of Las Vegas. The name Area 51 is the reference number of the box where it can be found on the map.

Knowledge of the universe

First of all the educational system can prepare people for society by teaching subjects like history, civics and the new subject concerning the theory about the universe, which I have already mentioned.

During the history lessons the history of the Earth, mankind and the different societies can be explained.

In the civics lessons the focus can be placed on what is happening in today's society. It is important that people start to understand why we are interacting in a certain way with each other and what we want to achieve by doing that.

During the lessons concerning the theory about the universe, the latest findings about the universe are being taught, including the knowledge of key documents such as the Torah,

the Tenach, the Koran, the Bible and writings of other major religions such as Buddhism and Hinduism. By extracting important teachings from these documents we will be able to learn and develop faster than ever before.

Another important part of the lessons on the universe will be spent on tools like meditation to implement the development of spiritual skills with which our consciousness can be deepened and anyone can get a greater sense of everything around us, including the entire universe. Through the acquisition of such knowledge and skills fulfilment is being given to the spiritual education of children and adults. This is an essential element in the spiritual society, but I do not think it is necessary to go into this any further. There is already enough knowledge available about these aspects to give completion in a proper and responsible manner.

What is best for the child

Another important aspect concerning the education of children that I like to stress deals with the individual abilities and needs of children. Since there are no more budgets for education,

there are many more opportunities to give contribution to this aspect. Moreover it is important that education, like family, is a reflection of society. If we want to see people living harmoniously together in society, taking into account each unique individual and interacting with each other in a loving way, then we have to start doing that in our families and education as well. This means that the groups at school should be much smaller. A maximum of 10 to 15 pupils per teacher is enough. And maybe there will be no more classical education after all. The teacher and its students are interacting with each other through webcams and are only meeting each other during exams. Who knows?

Every child should have the opportunity to follow an individual training program, tailored to his or her capabilities. Every child must be given the chance and the choice to do the things he or she likes to do, and to find out which abilities suit them best. If a child wants to know more about the fire brigade or about being a doctor, they should have the chance to make a visit there during the lessons for example. Finally, it should be made clear that the teacher is more like a guide than a leader who helps the child to find his or her way and place as a young adult in society.

In short, education will have to change. Another Society also means another way of educating children.

5.33 Summarizing

What more can I say about the spiritual society? A society with clear principles, objectives, visions and policies. A society in which love, happiness, freedom, personal development, and welfare of the people come first.

A society that is built on a system where the current known problems are resolved and in which the positive and good in the people is strengthened. A society in which there is no longer lust for power, greed, poverty, oppression, overpopulation, crime, economic crises and unmanageable environmental problems.

For whom this still is not entirely clear, here is one last very brief explanation:

1. no acts of power anymore, but democratic governance (see section 5.3, 5.8 and 5.9);
2. love, happiness and freedom as principles (see section 5.1, 5.3, 5.8 and 5.9);
3. abolish money, recording hours worked (see section 5.7 and 5.12 to 5.15);
4. equal opportunities and (development) possibilities (see section 5.1, 5.3 and 5.15);
5. education at home and in school as tools (see section 5.6 and 5.32);
6. balanced prosperity and welfare (see section 5.3, 5.4, 5.8, 5.12, 5.16, 5.20, 5.28 and 5.30);
7. detachment (see section 5.4, 5.7, 5.14 and 5.20).

A society that wants to be in balance with the Earth and everything that lives on it. Another important issue is that this system ensures that people will look for permanent and structural solutions for future problems, a society where we can enjoy a continued high level of prosperity and welfare. Where people have time for their children and for themselves. Where people are helping each other and where we take into account all life on Earth. A world where love connects us all with each other into eternity.

Doesn't that sound great? Isn't this what we all want? It can happen and it is really not that complicated. It is all about making the right choice. It is still not too late!

Chapter 6

Concept of the migration route

6.1 Introduction

I hope that when you have reached this chapter, you have become excited about the spiritual society. In my eyes the spiritual society is a beautiful next step, which we as humanity can make. I really think that this is the right time to go for it. The spiritual society will be built on centuries of knowledge, which have been retained through the years. In addition, using what we have learned from all previous and current social systems. Welfare and prosperity for all people will be at a high level, creating opportunities for everyone in spiritual enrichment, a society where everyone can live happy and healthy. A society where we believe that love and support for each other enable us to flourish, not only in this life, but also thereafter. This is a logical step to me. In any case, I become enthusiastic time and time again when I think what such a society could be like.

In some miraculous way I have had the opportunity to enjoy a wonderful life for many decades. I hope that may last for a long time and I wish everyone such a life. The fact that I am happy and others are not really makes me sad sometimes. I will not accept that happiness is not meant for everyone in the world. Maybe if I try very hard, I can make a few people happy near me or let them share my happiness. But actually I want everyone to be happy. That is why I wrote this book. And I believe that if we all want this, that it is possible, step by step. We have a choice!

The transition to the new society will be a huge step, because people should be prepared for this new society. The spiritual society requires a different way of thinking, another way of interacting with each other. This does not happen overnight. People have to integrate this into their way of thinking; others have to grow up with it. So much is going to change. Most people will not be able to picture this, perhaps not even understand. This could very well be the largest project ever launched. But isn't it worth saving our world? In the spiritual society we do not throw away what we have built, but rather we use it to build upon! Think of it as a huge step forward from our current society, a continuation in a modified form.

During the education, in schools and by training adults, everyone must follow the principles and rules of the new society. To let the transition take place in a universally acceptable way, a gradual transition is necessary. It is especially important to realize that if we do not transform this place everyone might lose everything. It has been clear for some time that we cannot continue with today's society in this way. If we lose control over the situation, we are much further from home. Probably many millions, perhaps billions of people, will lose their lives.

How can we save our world?

Moving on to the spiritual society is my answer, but that is something you as reader will surely not have missed. In addition, we can only save our world when all countries in the world are finally going to transfer. If only one country is going to transfer and the rest is still abusing the Earth, you will also be able to understand that the world, including ourselves, cannot be saved, even though we live in a country with a spiritual society.

The best way to proceed to the spiritual society is with all countries in the world simultaneously. When this happens there will be no need to maintain subsystems in order to enable the interaction between countries with a different type of society. The bottlenecks will lie on a completely different level, because there will be an enormous demand for products from all over the world at the same time, while the current production capacity is only sufficient for the rich part of the world. This requires a well thought-out plan. Things cannot change overnight. Put a number of wise men and women from around the world in one room and they will come up with a solution. Explain the plans accordingly to the United Nations and let them vote on it. When the plans are adopted, we can start straight away and our world can be saved!

Possible migration steps

This chapter will broadly describe the steps that need to be taken to gradually shift to the spiritual society. The assumption has been made that one country (or a few countries) will go first, because this is the most complex situation. This can also serve as a starting point for a plan, if all countries might decide to implement the spiritual society simultaneously after all. Not that it is simple, because this means that everyone should join, whether they like it or not. The risk of resistance from within is greater in this situation. Each movement usually leads to a counter movement. In this way the road in the middle is being created. That does not have to be bad at all, but you must be prepared for it. If all countries go simultaneously, then all countries will be simultaneously using the same system, making the whole world actually look like one country, without any another country beyond. The exchange of commodities and products will only be a matter of making agreements.

When money is removed from everywhere at the same time, people will have no choice, there will be no refuge. There is only the probability of an underground crime network to continue the trade in drugs and weapons, but this will only be of a short duration, because money laundering will be no longer possible.

Another advantage is that we will be able to assist developing countries much more efficiently. These countries will enter into a healthier situation faster. For the first time ever it is possible to start making realistic plans to help the developing countries in a humanitarian way, which will ensure everlasting progress.

Most importantly, the participation of almost all the countries in the world is necessary in order to save humanity and the Earth, as we know it today. The chain is no stronger than the weakest link!

Most of all look at the above as a possible scenario, and not as the migration scenario. It is a first version of a concept scenario. Undoubtedly, there are more aspects to be considered and there can be more phases to distinguish. Further details are necessary. This scenario is purely aimed to stimulate people, who want to join the making of plans. Undoubtedly there are other or better scenarios. For each country, a modified version needs to be applied. I cannot survey everything on my own. Through this plan, I would like to show which things we should think of for example, when we really want to transfer. In addition, I hope this migration plan will convince people that a transition to the spiritual society is possible. That it is not an unattainable utopia. I believe that if you really want to achieve something, you will. Set yourself a goal and go for it!

6.2 Step 0: making plans

Strategy → Plans → Executing manuals

The first step to realize something is always making plans, before you are really going to start. For a more accurate description: the first step should be the creation of a migration strategy. This is then elaborated in a number of migration (phase) plans. Each (phase) plan covers a different period in time and/or a different part of the migration subject. The strategy can be seen as an overall preparation, the plans as "what are we going to do?" and the executing manuals as "how are we going to do it?".

Since my main objective is to show that a migration is possible, I will not execute this three step planning scheme, but only describe a possible global scenario.

Design of the spiritual society

In the previous chapter, I have described what a spiritual society could possibly look like. In several areas a different or more specific description may be required. And perhaps there are areas that still need to be developed. It is important that all possible power factors are being neutralized or limited to a minimum. Love gets little opportunity, where there is power. Perform a review[1] of the design of your new society and adjust it.

When the design is approved, the migration strategy can be developed. It is important to determine which conditions must be met and what the starting points are. Only then is it known what should be taken into account. This will be helpful in making the plans.

Time to execute

For step 0, the planning phase, we should consider an execution time of about one year. A year seems a lot, but for a monumental undertaking such as this, a good preparation is necessary. One year is soon over.

Making plans

What needs to be executed and in what order, should be determined first at the start of making plans. It is therefore important to analyze the current situ-

1 Allow a number of wise ladies and gentlemen to read the design, discuss / review and make improvements, to develop a design that is acceptable to the majority of the population.

ation and to take a good look at the areas in which the transition to the spiritual society will have a major impact. Countermeasures need to be put into place to convince people in these matters. Many people are afraid to lose what they have accumulated or are simply afraid of change. It is therefore necessary to show that everyone's interests have been taken into account and that everyone is going to be better off. This must be included in the implementation plan. The implementation plan is, in addition to the adjusted design, the most important aspect of making plans.

When this has been done the steps to implement the spiritual society will be described.

When I look at a country like the Netherlands, I would like to propose the following steps (also see scheme 3):

1. national referendum
2. reform of science and religion
3. education and knowledge transfer
4. mid-term evaluation and Go / No Go - decision
5. political and organizational redesign
6. abolishing money
7. evaluating and adjusting

Ad1. A national referendum is needed to convince the majority of the Dutch people that this design and this implementation plan are the way to go. People need to be informed about these plans before voting can take place. The vote is expected to be positive as everyone will gain in the spiritual society. Executing time: about one year until the time of holding the referendum;

Ad2. The next step is creating a foundation for the spiritual community. This is done by providing the required knowledge about the universe, worked out in cooperation with science. In addition, different organizations, like the religious communities, churches and science must be prepared for their new role in spreading the new knowledge.
For this research phase and coordination between various parties, at least two years executing time seems reasonable;

Ad3. Then the new foundation for the society must be promoted at large scale and taught to the people. Execution time: about two years;

Ad4. Since the previous steps are not an irreversible process, but essential to be able to take the next step in the migration process, this is a good time for an interim evaluation. Leading on from this it can be decided whether it is a Go or a No Go. Executing time: about one year;

Ad5. Now is the time to change the organization of the society in preparation for the start of the new process that underlies the spiritual society. Because of the size and the enormous impact of this step the executing time is difficult to estimate. It should be between two and five years;

Ad6. When the previous step is completed, the new society process can start. The decisive factor is the abolition of money. That is the decisive moment: is the spiritual society really going to work? For this step I would suggest a period of two years, but perhaps it is possible to combine the preparations with step 5, so that this step can be shorter in duration.

Ad7. After a short while, a review should take place to identify bottlenecks, to rectify them and possibly adjust one or several basic principles of the spiritual society. When this step has been executed successfully, the spiritual society has become reality.

For a first evaluation phase not much time should be planned, because major bottlenecks must be resolved quickly in order to maintain the motivation at a high level. That is why I would suggest a period of one year maximum.

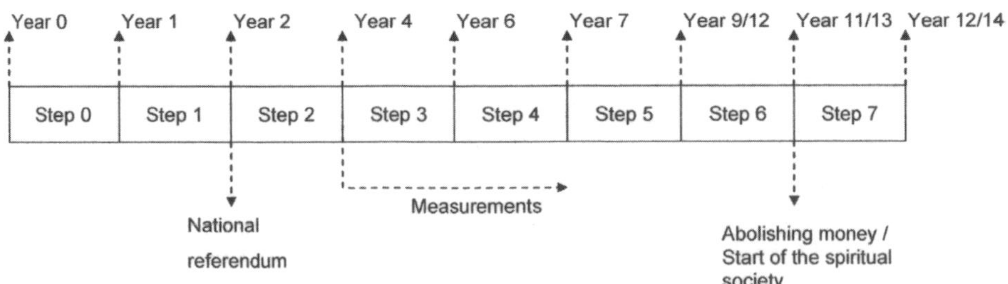

Scheme 3. Technically this is what the planning of the roadmap looks like.

Notes:

1. In particular, step 5, the political and organizational redesign, will take the most time. For this purpose I have taken into account a period of 2 to 5 years to get this arranged;
2. The spiritual society will only really begin when money has been abolished
3. From the beginning of step 3, measurements will be in place to be able to make a Go / No Go – decision at the end of step 4.

In the following paragraphs, this plan is described in more detail.

6.3 Step 1: national referendum

The first major step is to convince the population that the spiritual society is a solution to virtually all major problems in the world and the most loving way to a balanced and happy life for everyone. The decision should be taken in a national referendum. (unless all countries in the world are going into it simultaneously, then the decision will be taken in the United Nations)

Informing and promoting

In preparation of the referendum a broad campaign should be launched. Government representatives should travel through the country to inform people about the spiritual society and to allow them to ask questions. Through television spots, discussion programs and documentaries, the people need to be informed about what the spiritual society has to offer.

Seeing the benefits

Ultimately everyone will benefit from the spiritual society, but a small group will probably not see this immediately. Most people will see the benefits immediately. They have few possessions and need to work hard for a somewhat pleasant life. In the spiritual society, they will certainly get a better life. The average level of prosperity of these people will go up significantly. In addition, they will get more time to work on their own development. But the most important thing is the fact that they will get the opportunity to do something that they really like, something they have chosen for themselves. Not something that they accidentally rolled into and where they have the feeling that they won't ever escape. Moreover, they are supported in this from the society.

The group that will probably not see the benefits of these changes immediately are the rich and powerful in the current society. As far as their property is concerned, they will hardly notice any change. The bottleneck lies in the abolition of money and power. Of course, they remain in the government to rule the country and company owners will continue to manage their businesses, because it is important to have the right people in these positions. But they lose a piece of power over the people, because employees won't be as dependant on employers when they are searching for a suitable job., The employers will need to do their best to make things enjoyable for their employees, more than previously. Ultimately it will be better for the atmosphere and the development of people, because everyone will give their honest opinions. People will start to be themselves.

The rich people will get the feeling that they have no control over what they can buy anymore, due to the abolishment of money. That is absolutely not true. Absurdities, such as large quantities of petrol and diesel consuming automobiles are no longer allowed. They may be retained, but no longer used. You cannot consume unlimited amounts of anything. Everyone is treated equally in this case. This means that one person will have to do with less than before while the other will gain something. But enough is enough, isn't it?

Moreover, the acquisition of wealth in today's society is only possible because the system that we know, allows this. If we had had a different system, maybe there wouldn't be such a thing as rich person. Where does this wealth come from? Some of the rich people belong to the old aristocracy or old money. That money comes from the past and has become their possession through inheritance. The following comments can be made on being rich:

A. that money is often obtained as a result of power over other people such as slavery, war, child labor, from colonies, plantations where slaves worked, big landowners who oppressed farmers by being cunning and deceptive;
B. a lot of money that has been earned in a fair way, if it hasn't been a matter of sheer luck: sufficient intelligence, the right parents, doing something at the right time, outer beauty, special talents, etc. Let us be honest: good ideas and hard work will make only a few people very rich;
C. money usually leads to power. 'Power over' often implies oppression and causes violence and war. Power is a dangerous and frightening thing. There are too few people who can handle this in a responsible way.

In short, to be rich is not something you should be proud of in relation to other people. It is debatable whether it is right and fair to let this situation exist the way it is.

The big opportunity

In a way I can imagine the views of the rich and powerful. But isn't this actually nothing other than reasonable? In the current situation, the ordinary people, the vast majority must adjust themselves all the time and in the spiritual society the rich and powerful must adjust themselves, just a small minority of the population. This is the opportunity for the rich and powerful to show that they really want the best for their fellow human beings. This goes further than just giving money to charity. This means really giving up something for charity. The difference with giving money to charities is that the rich and powerful are really improving themselves by contributing to the spiritual society, only in a more mental and spiritual sense. Isn't that more valuable than a large bank account? Pretending to have done something good by giving a very small part of your money to a charity is no longer necessary. Everyone has the opportunity to do something really good by simply doing it. Since money no longer exists, the amount is no longer relevant. It is what you really do and mean for your fellow man that is important.

The rich people have to deal with fraud, theft, burglary and other crime. These problems will be solved in the spiritual society. But the most important thing is that the rich and powerful people can lose a lot, when they are prepared to risk everything by endangering our world further. Undoubtedly, not all rich and powerful will come out well when the Earth starts to respond to the disruption of the natural balance. Everyone who risks a lot can also expect to be confronted with significant losses.

It is best to think of this change as an investment in the future, a future that will not only bring prosperity for everyone, but above all mental and spiritual wealth.

The result of the votes

It is in the line of expectation that a vast majority of the population will vote in favor of the spiritual society. Partly because almost everyone will be better off and partly, because a large part of the population will see that every effort should be made to save life on Earth, because without life there is no future whether you are rich, powerful or poor.

6.4 Step 2: reform of science and religion

Spirituality, love and meaning are important principles of the new society. Science and religion play an important role. They will have to work together to shape these principles. People must get the right perspective.

6.4.1 Religion

The various religions nowadays focus on the details in the religious scriptures. This created fierce discussions and contradictions. This is what I call the negative approach. If we look at the headlines of all religions, then they are almost the same. In addition, events like near-death experiences and the paranormal gifted teach us that there is a link. All these agreements and outlines should be described in a new universal writing that will be given to everyone, in their own language. This is done in step 3, education and knowledge transfer. This is the positive approach to faith. The rules are so simple, the description clear and brief, that there can be no power derived from knowledge of the writing.

Collaboration
An important additional step, however, is to agree with the world religions that they play a role with regard to step 3. The leaders of these religions should be convinced that this is the way that the world should go. That love, solidarity and cooperation is the only solution to humanity's survival, and that they must set aside their own interests for the interests of all people. They will have to prepare a program to tell their followers about the new society and to guide them on this spiritual path.

Basis for the society
Because many leaders will want to maintain their dominance and many believers are convinced of their own religion it is important to emphasize that it is not the intention to deprive people of their faith, but that it is all about a new basis for the society. Together with the scientific support, this should also work positively for almost all religions.

6.4.2 Science

Science is too focused on 'proven' facts, which causes science to fall behind on reality. Before something is scientifically proven, something must have occurred repeatedly. Then there are all sorts of calculations that need to be made to be able to leave out all kinds of relations. Smoking is a good example. Actually we already knew for decades that smoking is harmful and may cause lung cancer. But it was only when it was shown statistically that many people who smoke die from lung cancer, that the fact was officially recognized. Suppose we do the same with the relationship between the quality of life on Earth and our current way of living: just when there is no quality of life anymore, we will conclude that there finally is scientific evidence? I think that we are too late then. How many people still have to die from cancer in prosperous countries, before we start to see a link? Must the expectation of life go down first? This will certainly happen in the coming decades, but do we want this? Science should take a more proactive attitude. That means taking the lead in the developments within the society, gathering facts about the things that are a concern to the community, and preventing developments escalating, before it is too late.

Society supporting science

What we need during the migration to the spiritual society is knowledge about the universe, near-death experiences and paranormal phenomena. Carl Gustav Jung's theory about the collective unconscious is being considered as science in psychology. Isn't it time to do some further research to learn more about this subject? Jung talks about a collective unconscious, which includes basic knowledge. This knowledge is being transferred from generation to generation and is increasingly expanding with knowledge of general use to mankind. Thus this helps mankind to develop further. It is not known how this transfer of knowledge takes place. On the other hand, when you believe in reincarnation, the concept of the collective unconscious is once again quite understandable. There are also studies on near-death experiences, which appear to indicate that consciousness can exist outside the physical body. This also fits into the picture of the collective unconscious. It is the art of science to detect and prove these relationships. So long as each science only focuses on statistical data within its own area and does not look beyond this, we will not progress much further.

It is also debatable whether certain relations can be proved with the current scientific methods. For example, how can we prove something that has no physical existence and that cannot be reached with physical equipment? Doesn't it exist or have we now reached the border between proving and believing? Science is still looking for the end of the universe. We are now able to look around us up to a distance of billions of light years. According to science nothing further exists, because it has not yet been proven. But that does not mean that there is nothing, try to understand that! The reverse is not true: everything you can think of is true, until the contrary has been proven. By some logical thinking a lot of things, which people have made up, can simply be explained.

On the other hand, many new facts can be proven and conclusions can be drawn based on existing information. In that respect we need to shift from the idea of 'science for science' and move more towards a society supporting science. What science is doing nowadays is usually for a particular financial interest, but we want to do this in the spiritual society to help the people. Thus, knowledge of the functioning of the universe will open the eyes of many people and lead society in the right direction. The found "scientific" knowledge should be added to the universal writing from the previous paragraph.

Another important contribution that science can provide is the design of educational and teaching programs to integrate this knowledge within the society in a responsible way. It is also important to explain to people, based on scientific data and analysis, why the current way of living can no longer be maintained.

6.4.3 Uniting the world

The knowledge enshrined in the universal writing, including the fact that love is what binds us all together, is of essential importance. This should be the foundation of our society. This knowledge will bring the believers and non-believers together. This knowledge will unite the world.

6.5 Step 3: education and knowledge transfer

Educational program

When step 2 has been completed, the (re)education of the current and new world residents can begin. For the current world population it is important to know what is going to change. They will have to democratically accept this and be willing to transfer this to their children. Special events and educational workshops will be created for them, bringing fresh knowledge. Furthermore it is a place where they can learn what their role is as far as educating their children goes, and what they should teach them.

Universe study book

The book, developed in step 2 with universal knowledge, will be disseminated among the population. The content will be taught during workshops and people can explain where the ideas come from and why they are important. Furthermore, the content of the book will be used in educational (television) programs. Even if people do not read the book, the knowledge will be disseminated among the population. You can read the book yourself, but you can also use it as a reference.

Cooperation of the media

We also need to clarify what steps we as a society are going to make. This can also be used in documentaries and feature films, in which results from different scenarios are shown. In this respect the media and entertainment will play a very useful role and can demonstrate that they can do more than just entertain people.

Discussion forum

However when an opinion is being expressed, there will always be people who think differently. People use the form of an argument to express this. This cannot be prevented, even if it is about saving humanity and the world we live in and even though everyone will be better off, because not everyone will believe this. Or maybe they are afraid of abuse, change or that they may get in a worse position. And maybe they are right.

Therefore it is important that this change is supported by a democratic process, in which the choice of the majority of the population is being followed. Opponents to this change should also be heard and be given the opportunity to put their arguments forward. Discussion forums on television can show that these arguments can be rebutted and thus contribute to the positive image of the spiritual society. I believe that a positive approach to people can lead to success of the spiritual society, in accordance with the principles of this society.

Timing
We will be ready for the next step when we have convinced ourselves that there has been sufficient knowledge transferred and integrated into the human behavior. This step should not be taken lightly. The duration of this step should not be too short, because then the population will not be ready, on the other hand it should not be too long either, because then people will get impatient and lose faith in the spiritual society ever happening. The remaining negative opponents will seize this time to prevent the continuation of the start of the spiritual society. This period is therefore crucial for the next phase.

6.6 Step 4: intermediate evaluation and Go / No Go-decision

Evaluation
The central government will evaluate the situation in society in order to see whether the knowledge about the spiritual society is sufficient and noticeably present in the changing society. This can be done by looking at statistics, is crime decreasing (are there fewer bicycles stolen, fewer burglaries fewer attacks carried out, etcetera), are people making less mileage, are fewer people being fired, are there fewer reports of discrimination? This can be done by doing surveys, analyzing evaluation reports of the educational programs, by examining the hours spent on education, but also by real live observations by police officers and city guards.

Go / No Go-meeting
On the basis of these studies, after about 3 years a Go / No Go-meeting should take place to decide whether the present society is ready for a political and organizational redesign. This is only possible if people have a different mindset than the current and are already thinking somewhat in the atmosphere of the spiritual society. The impact of the political and organizational redesign is very large and acceptance is essential for this step to be a success.

Preparation of the political and organizational redesign
There will be major changes, especially in the way the state is being governed and economically organized. It is important that there is enough consideration regarding people's lives and feelings. This is the first moment at which we can show that the spiritual society is taking people into account in an effective manner. The fact that things are changing is not such a big problem

for people, but by doing it in a people-friendly way; you have the chance to strengthen the belief in the ideas behind the spiritual society. These opportunities must not be missed. It should be noted in advance that the plans must comply with this aspect, to minimize the likelihood of problems.

6.7 Step 5: the political and organizational redesign

6.7.1 Governance

Change managers

The next step is to establish the organization of this new society. Many old-fashioned politicians, managers and directors will not go along with these developments and must leave their jobs. Old values such as money, power and ownership are going to disappear over time. If leaders here want to continue to attach importance to it, they will have to seek another place on Earth where they can live in that way. It is important that people with a positive state of mind and a vision towards the spiritual society are placed in key positions to shape the organization spiritually. Managers should set an example and reform the rules in society and businesses in a way that will meet the principles of the spiritual society. Actually, they have to ensure that a spiritual society becomes reality, but then using money for payments. This is a difficult aspect, because as long as there is money, the negative effects will also remain. These could impose on the current movement towards the spiritual society. This must be avoided at all times. The new managers should therefore act firmly and be real managers in change.

Expropriation of companies

The main element in this phase is the "expropriation" of companies. This means that companies only need to be managed, but are no longer personal property. The residents of all the houses in the country are given permanent ownership. Mortgage debts are forgiven and are discarded. The government is only going to pay back debts, when they are held by foreign companies. Companies, that no longer have a reason of existence as a result of the spiritual society, will come to an end and new jobs will be found for the employees. It is important that no one gets into trouble while this change is taking place. This is a responsible task of the change managers, who must accompany the redesign.

Insurances and taxes

Another important task is to dismantle the insurance and tax system. All insurances must be eliminated in this step, as well as the social security taxes and the taxes for inhabitants. All insurance cases will be paid and handled by the government authorities from this moment on. The expenses of the government will significantly increase. On the other hand, the competitiveness of domestic companies will increase strongly, making them more profitable and therefore revenue of taxes from businesses. The corporation tax percentage will be significantly increased, to 50% for example, because there are no more shareholders, they do not have to pay dividend. This saves the companies a lot of money.

Customized employment

Everyone should think about his or her job in relation to the reorganization. Is this the work I want to do or am I going to do something that I really enjoy doing? Many jobs in businesses such as the financial service delivery and the tax offices will come to an end. But in addition there will be many new jobs in the production, construction, agriculture, horticulture, home furnishings, care sector, developing countries, etc. This is the time to look for something new, something that you really like or always wanted to do. And after the reorganization there will be many possibilities for each person to develop, without a negative effect on their personal situation or home life.

6.7.2 Companies

Continuity

All companies in the country can continue to run, as they are now. But there is a difference between foreign-owned firms and those in domestic hands, and between companies with sales in this country and exporting companies. Because they are still working with money abroad, we must ensure that there is a continuous flow of funds to keep our own country running in the world. For example, the Netherlands has become a country that delivers services, where many goods are imported. As long as these come from countries without a spiritual society, they must still be paid.

Companies in foreign hands

Companies in foreign hands can stay for a while. These companies will keep paying taxes to the government. This money can be used for the import of goods. In addition, they are given the opportunity to sell the company to the government department. At this stage these companies will find it increasingly difficult to get employees, because the culture of these companies is different from the spiritual society. In the next phase, these foreign companies will disappear, because they can no longer earn money in this country. The only possibility for foreign companies, who make products here, would be to sell to countries that are not in the spiritual society (as described in step 6) or to sell to foreign firms from countries that are simultaneously introducing a spiritual society.

Domestic companies

Domestic companies will lose their shareholders at this stage. All shares in foreign hands will be bought by the government. Trading shares on the stock market will be stopped immediately. Domestic shares lose their value. The current owners of companies are maintained. The emphasis is not on ownership but on managing the business. Only being the owner and not a manager is therefore a problem. People must be contributing to the spiritual society to have a right to services and products. It is important that the company culture is being migrated to a culture that is appropriate to the spiritual society. This is an important task for the managers' new-style. There will be little change for companies which are focused on the domestic market.

Export and competition

New businesses will be established, which will start to export products. In this phase, the labor costs will be greatly reduced (see the sections on money flows). This makes production much cheaper, which will strengthen the market position in relation to foreign companies. In the next phase the domestic financial system will be abolished. Producing products will hardly cost anything, so most of the sales will be profit. We will then be able to compete with almost any country in the world.

In this step, the production will still be low in some sectors, because we are in the start-up phase. The revenues are not going to the government department at this stage, but are transferred to a separate business account. The money is again used to import products and raw materials that companies need to produce and/or sell and to pay the workers. Due to the increasing profit, there is also more tax paid to the government. This reduces the government deficits.

6.7.3 Homes

Homes and land in foreign hands

All houses and land in foreign hands are bought by the government authorities. Transactions in houses are immediately put to a stop. Everyone in this country, who owns a house and land, may retain that and cannot sell it. Exchanging is always allowed.

When people want to emigrate, their possessions will be valued. The amount of money will then be transferred to a foreign account by the government authorities.

The right to land and housing

Every adult has the right to about 250 square meters of land, so each couple or family has 500 square meters land. Every single person or couple is entitled to approximately 350 cubic meters of residential contents, and each family with 2 children has 500 cubic meters of contents. For every extra child up to 50 cubic meters may be added. This does not include a barn or garage. People, who are now living in a smaller house, may as a result choose a larger house from the vacant homes. The people, who are living in a larger house and/or a large piece of land, may continue to live there. The land will be later redistributed and rebuilt on when the last residents leave the building, unless of course it is not appropriate to do so for historical or scenic reasons. Most of these estates will be later used as offices, accommodation, shared housing or for museums. Thereafter more people will be able to live in a responsible manner, in a beautiful place.

6.7.4 Domestic money flows

All insurances through the public government

Insurances should be abolished. If something happens, the state will be financially liable.

The government will be confronted with a large debt, but this is only for a short period of time. The lower production costs of companies will ensure a trade surplus, causing the debt to decline quickly. The increased corporation tax for companies will also contribute to this.

Salary only net
The salary will be net for everyone and about half of the current gross income. In addition, most of the fixed costs will be dropped: pension, mortgage, rent, social security, taxes and insurances. You still have to pay for energy, costs for food and other products in stores, because much of this will continue to come from abroad and the people in the companies must still be paid in this phase.

Intermediate phase
We must realize that this is an intermediate phase. The money will only be maintained, because there are many elements to take care of, which will prevent us from moving to the next step. The abolition of money is a big step. As far as domestic money flows are concerned, the most important thing is to make sure that no one gets into trouble. The welfare level is already slightly leveled. A small group of people will have to get used to this. It is therefore very important that Step 3 has been successfully completed, especially among the rich and powerful. It is important that they understand that they have to set an example and that they play an important role in saving our world. Self-enrichment and pollution are not good examples for a better and more balanced society. That should be clear to everyone in this phase.

6.7.5 Foreign money flows

Transferring money from foreign accounts
The only reason that we have maintained the money is because we will (always) partly continue to depend on foreign countries. As long as these countries do not have a spiritual society, we must use money as currency. Therefore we will need to ensure that enough money is available to import what we need. All foreign banks will be contacted in this phase and asked to transfer money from accounts of all the inhabitants of the upcoming spiritual society to accounts of domestic commercial banks. The banks will be informed when the money of the accounts must be transferred to the government account in the next phase. In the next phase this money is required to be able to establish a strong government, which will ensure the import of all necessary products and the financial money flows with foreign countries.

Commercial banks

To regulate the money flows with foreign countries we need a separate independent authority. The commercial banks will get this function. The money from the government and the 'old' companies will be managed by a limited number of independent commercial banks. The objectives being that they are going to regulate the money flows with foreign countries of imports and exports in the name of the government. These money flows must take place independently from the money within the country. The banks which cannot or do not want to participate, will have to leave to a foreign country. The banks that remain are in fact going to be 'state-owned' banks. The staff of the banks, like other citizens, will be paid by the state. The government authorities will perform checks on the ins and outs of the commercial banks. In the Netherlands this will be done by "De Nederlandsche Bank" (DNB).

Conclusion

The domestic and foreign money flows are coming together in the government budget. It is important that no deficits remain in the long term.

6.7.6 Energy

Energy is an especially important necessity. Therefore we must take extra measures to ensure that our energy supply is independent from foreign countries. Think of solar panels, windmills, watermills and so on. Creativity is a very important factor for our independence. Moreover, a high energy production could mean exportation to abroad, which gives us means to import products and raw materials.

We will therefore have to focus on reducing our energy consumption and the building of energy plants, independently from abroad, for a number of years. We can only proceed to the next step when we have our energy consumption and production in balance. The aim should be to produce clean energy, because one of the causes of our unhealthy society is the pollution of the environment.

6.7.6 Nutrition

Another important need that we humans have is nutrition. We will have to learn to provide ourselves more with healthy nutrients. This means that the agriculture and livestock production must be reorganized. The emphasis should be placed on healthy, organic sound production and not on mass production. Nowadays production is heavily based on stimulants, but they are causing more waste than the environment can handle. Cows for example are growing faster and are producing more milk than ever. And also more waste than ever, which the land and water cannot process fast enough.

If we eat healthier there is less chance of contracting diseases in the long term. This is only possible when we are responsible for the production, and the transporting lines are short, so that the amount of preservatives can be minimized. The more people live healthier, the better they will function.

6.7.7 Execution time

These reforms will take around 2 to 5 years. In the first three steps the people are made aware of the movement towards the new society. In this step, the first major social changes will be established. This phase should not take too long, because there is still a form of financial economy, where income differences are still being maintained to a large extent.

An important aspect is the movement to a greater independence of other countries and the preparation of the society for the abolition of money. This should be done carefully to prevent major financial problems for our country, which could slow the movement to the next step.

6.8 Step 6: the abolition of the money

Competition with foreign countries

Furthermore money will be abolished, although not completely because we will still need to be able to handle the competition with foreign countries in a effective and efficient manner. Our production will no longer be faced with costs, which will ensure our products are cheaper than in the capitalist countries. Exports from our country will be significant. The money that it is earned goes to the company accounts in commercial banks and will be used for the import of raw materials. The rest of the profits will go to the state, which will then buy products, which we do not have or are not able to make in our own country.

Foreign companies

The companies no longer have to pay money for water and energy and also do not have to pay salaries to employees anymore. This will cause their earnings from abroad to rapidly increase. Our companies will set an example for foreign companies.

The foreign companies in our country will all disappear at this stage, because the profits must be fully paid to the government. Therefore there is no money to be earned here.

Financial economics versus 'economy of happiness'

Money will no longer play a role in our own country, but the economy still does. We will have a financial economy in relation to other countries. Our domestic economy can best be described as an 'economy of happiness'. From now on the most important factor in our country is your happiness. All the material aspects have become a sideshow. People will have to get used to the new situation, because how does it work when you step into a shop to buy something? All bank cards will be replaced by ID-cards linked to a payment system based on the number of hours you have worked. Every adult will receive an ID-card containing the authorizations, which apply to him or her. These cards will also indicate whether you are allowed to 'buy' products. Money no longer plays a role, so the products are rapidly going to change. Some products will be sold out quickly, while others will stay in the stores. This calls for a refocus on the products. The same goes for houses. The small and narrow houses will become empty. Houses will no longer have a value, because there is no money anymore. There is no longer a price card attached to it. When you have a job, you can choose a piece of land and a house that you want. There must be limits in place, because otherwise there might not be enough ground. The houses that become vacant will be checked to see if they can make a suitable location for someone. For example we can make houses for young single people, immigrants, or the elderly who do not want a large house with high maintenance and a large garden. The houses and flats that are permanently empty will be demolished and new houses will be built to meet the requirements of the people. If several people want to live in the same place, a fair lottery will be organized under the supervision of the notary.

Money transfer to the state

All the money that people have on private accounts at commercial banks will be transferred to the state and accounts are eliminated. Also all the money from insurance companies, banks, investment companies, stock trading and pension companies will be owned by the state. All resulting deficits from the past are being resolved in this way.

The resulting surplus can be used for trade with non-spiritual societies. The government fulfils a supporting role.

Identity card (ID-card)

The number of hours worked is administrated in a personal file of the state. Each resident receives such a file and an ID-card. These cards are issued from birth and returned at death. What if the person does not work enough hours? These hours can be caught up in the following week or month. If that person does not succeed in doing this, contact will be initiated.

Everyone can get a job that he or she likes, provided that the necessary skills are there and the required training has successfully been followed, this will enable them to live where and in what kind of house they choose, in addition to buy whatever they wish. The idea being that those who contribute to the society are entitled to this.

However, we will have to take additional account of each other at this stage, because this is a big step for our society. If things go wrong, due to the abolition, this should be reported to the government as soon as possible. A solution will be devised to prevent these problems from happening again in the future.

We must all realize that this is the path we want to go down to reach a better and fairer society.

The last 'old fashioned' countries

Eventually there will only be a few countries left where money can be used. At that time there will be enough countries in the world with a spiritual society to be able to completely shut down the money system. The remaining countries with a financial system will need to come up with a solution, if they want to continue doing business with countries with a spiritual society. The expectation is that these countries will experience this as a cumber stone and will switch to a spiritual society. At that time, polluting toys and excessive luxuries will no longer be in production, making the difference very small between the spiritual and the remaining countries. Most multinational companies have already withdrawn from the spiritual countries. There will be hardly any countries to which products can be sold for money, causing the profits to decrease significantly. Stocks and stock markets do not exist anymore. In short, the possibilities to enrich yourself with money will diminish when almost all countries migrate to a spiritual society.

6.9 Step 7: evaluating and adjusting

One last important step is to evaluate what has taken place and to look at the points that need to be adjusted. A society is and remains complex, whichever way you look at it. Each individual is just a little different from the other and different people have different requirements, desires and needs. To keep society running in this way for a long time attention to these requirements, wishes and needs are of major importance. Because the conversion is great, it may be the case that some things are less well implemented, than previously thought or certain aspects may not be considered enough. During this review, these cases should arise and be adjusted accordingly. Actually it is just a large project, only the budget and duration are unlimited. The quality of the final result must be higher than high. In this regard it is a project in the right proportions. The quality of the final result is unquestionable. Love and happiness for all should be achieved. Special teams should be set up for the adjustment tasks. They will make separate plans and will start to realize them. A good approach is prerequisite for a happy ending.

6.10 Conclusion

The spiritual society can be achieved

The spiritual society seems to be far away when you are reading chapter 5. After reading this chapter it becomes a lot nearer. And that is precisely the purpose of this chapter. It is already difficult to imagine another society, but the road to it is even more difficult to imagine. However, everything we can imagine is possible. In this case it is mainly a question of maintaining a positive attitude, to motivate each other to choose and to start making the step towards the spiritual society together.

Migration is a first step

The described migration has been a first step, no more than that. It is important that ultimately the entire path of 'from strategy to plans and from plans to execution steps' are made in order to ensure that there is a well thought off migration schedule to be able to minimize problems during this process.

Hopefully I have put a number of thought processes in motion with this first step. Maybe there are people and organizations that see the realistic possibilities and can do something with these ideas.

A multi-year planning is necessary

Finally, I want to point out that it will probably take more than 10 years for these plans to become reality. And in the meantime, we must not forget to start working on the existing problems in society, because otherwise things will only become worse in the meantime (see Chapter 2).

I do not know how many years we have left before the relatively stable situation of this moment is over and we end up in an irreversible downward spiral. If we wait too long we will no longer have the opportunity to realize the spiritual society. That would be unfortunate. Especially in the capitalist society we have learned that we should use every opportunity for progress. Let us not overlook this opportunity.

Chapter 7

Final word

Hopefully it has become sufficiently clear that all of the above is my vision, my image of the past, present and future. I would not suggest that all these facts to the smallest details are correct. I have named a series of facts in order to create an image. I have tried to show you an image of humanity through time and highlighted various aspects of the current society. I have explained the relations between these two, and extrapolated what followed. Firstly in the direction we are going and then in a direction we could go: the spiritual society.

My image of the society does not need to be your image. Everyone sees the reality differently. Maybe you believe that everything will be fine the way it is and you are not at all interested in a new society, as I have outlined. And if there are far more people who think that way, then something like the spiritual society will never happen.

But hopefully you are aware that you have a choice. That society as we know it, is not a fixed situation, but only persists as long as we opt for it. Every day we make that choice, consciously or unconsciously. We have a choice! Also when you are only participating in this society you make a choice. You may think you do not, but you do. As long as enough people contribute to this society, there will be little change. When we indicate that we want a better society, and that it is possible in a democratic way, then we can achieve this. There is a better society possible on this Earth, but only if enough people want it. Make your choice!

My vision for a possible new future is my attempt to generate ideas for a more loving, more honest and helpful world. A prosperous world, where people live together in freedom and prosperity. Where 'together' is no longer a concept from an economic standpoint, but mainly from a social point of view. A world where we live in balance with the Earth and every living creature. A world where people help each other, because the other person really needs help, not just because there is money involved.

It is not my intention to tell everyone literally what should happen. I just tried to outline a coherent whole that could possibly work. Large changes in my opinion can only work if those changes are supported by all groups in society and if all those groups are working together in a positive way to complete those changes.

On the other hand, this is an attempt to make everyone understand that this society and perhaps humanity are slowly coming to an end, as long as we stick to the current capitalist and communist systems, which are getting more and more like a dictatorship. I have cited a wide range of facts. I want to show everyone that the world is not as good as many of us would like to believe. For as long as we cannot see that things need to change, it will not make any sense to talk about change.

It was easy for me to line up a lot of the problems and shortcomings of the current world, as we know by now. And I am still far from complete. Man has its shortcomings, but that is no excuse for going on the way we do.

I hope that this book has left a positive impression and that I have shown that we have a choice, that there is hope for everyone on Earth, for all the lives that remain. Try to maintain that positive feeling and continue thinking about what you have just read. Tell others about it, talk with them, learn from each other and hold on! The more people that start to think and believe in a positive revolution, the greater the chance that it will eventually happen. This is not for me, but for yourself and especially for everybody who would like to live happily after us on this Earth.

In any case I am grateful to you for the time you've taken to read this book. This is a first step.

In this concluding word I would finally like everyone who reads this book to do the following:

'Give me a response by e-mail' (through www.spiritualsociety.nl)

All responses will be processed. A complementary will be made on the basis of this processing. The results will be made available through the Internet, so that any reader of this book has free access to this information.

And hopefully you will also pass this message on, so that the spiritual society is really going to happen and we can make a new step forward!

Annex

The following is more or less literally taken from the website of Bob Coppes, where he briefly talks about his book "Near-death Experiences and World religions".

Near-Death Experiences and World religions, Bob Coppes
In his book, Bob Coppes compares the essences of the NDEs with those of five world religions. His conclusion is that the real essences of the religions can be found in NDEs, but not all essences of the NDEs can be found again in each of the religions. Consequently, NDEs are more universal than each of the religions separately. (in Dutch: ISBN 90-5911-680-1)

Other main conclusions
According to Bob all NDEs give us an insight into what happens during the first moments after our death. Interesting is that people who have had a NDE seem to get an answer to the question 'what the purpose is of my life on Earth'. The experiences also seem to contain a message about how we should live, what we should do with our life and what we shouldn't do. Other conclusions from his research are:

1. We all belong to one big whole: we are part of a unity universe.
2. We are all a part of the Light (or God, Allah, Jahweh, Brahman or however you prefer to call it). "Each of us is like a grain of sand on the beach, while God is the whole beach".
3. The most important thing by far in this unity universe is Love. It seems to be the building block of everything.
4. Our consciousness continues to exist in this unity universe after the death of our physical body.
5. The Light does not punish. It is unconditional love, love without conditions.

In the book by Bob Coppes the foreword is written by, the now retired, cardiologist Pim van Lommel, who researched medical science for years after the near-death experiences of people who survived a cardiac arrest. The research of van Lommel was published in the leading scientific journal "The Lancet" in 2001 and was then front page news around the world. At the end of 2007 he wrote the book "Endless Consciousness", in which a scientific vision on near-death experiences is given in a readable manner for a wide public.

Back cover of the book

Many books have been written about how we might save life on Earth. Some people believe that saving the Earth will cost 125 billion dollars while others claim that 1000 billion dollars will not be enough to achieve that goal. The solution as described in this book will not cost a penny! Indeed, all people will get the chance to have a greater inner wisdom and material welfare. Forming a better future requires a different way of thinking, a different way of living together. That is the message of this book.

On scrutinizing the history of the world, one may discover a number of remarkable issues which are typical for human development. It determines the state of affairs in our present society and reveals a number of serious problems putting our entire world at risk.

This situation is giving rise to a number of possible future scenarios which do not seem to be all that bright and shining. Are we going to let this happen? Does mankind want to evolve like plants and animals or do we want to show that we are intelligent creatures determined to shape our own future?

Look at the positive developments in our current society and look at the results a spiritual society may offer. The spiritual society may provide different objectives and principles and might genuinely solve problems. Thus, it might be a world of love in which people strive to be in balance with our planet as well as with the universe. This ideal future might become a real world in which the term `together' refers to warm human values instead of purely economic ones.

Your future, our world?